MARKET SOCIA

MARKET SOCIALISM

EDITED BY
JULIAN LE GRAND AND SAUL ESTRIN

CLARENDON PRESS · OXFORD
1989

Oxford University Press, Walton Street, Oxford OX2 6DP
Oxford New York Toronto
Delhi Bombay Calcutta Madras Karachi
Petaling Jaya Singapore Hong Kong Tokyo
Nairobi Dar es Salaam Cape Town
Melbourne Auckland
and associated companies in
Berlin Ibadan

Oxford is a trade mark of Oxford University Press

Published in the United States
by Oxford University Press, New York

British Library Cataloguing in Publication Data
Market socialism
1. Socialism. Economic aspects
I. Le Grand, Julian   II. Estrin, Saul
335
ISBN 0–19–827701–6
ISBN 0–19–827700–8 (pbk.)

Library of Congress Cataloging-in-Publication Data
Market socialism / edited by Julian Le Grand and Saul Estrin.
Bibliography   Includes index.
1. Socialism.   2. Mixed economy.
I. Le Grand, Julian.   II. Estrin, Saul.
HX73.M345 1989   335.8—dc19   88–38793
ISBN 0–19–827701–6
ISBN 0–19–827700–8 (pbk.)

Printed and bound in
Great Britain by Biddles Ltd
Guildford and King's Lynn

Preface

AFTER Labour's defeat in the 1983 General Election—the second defeat in a row—the Fabian Society called together a group of sympathetic academics and others to a meeting to discuss what had gone wrong. At that meeting Julian Le Grand and Raymond Plant argued that perhaps the major current problem faced by the Left, including the Labour Party, was its loss of an intellectual base. Many of the traditional socialist forms of economic organization—such as central planning or nationalization—were widely perceived as failures; even collectivist values were discredited. The Left was no longer in the vanguard of intellectual radicalism; rather it was the so-called New Right that was producing radical ideas for social reform and change, ideas to which the Left could only respond with a limp defence of the status quo. What was needed was nothing less than a rethink of socialism: a re-evaluation of its basic tenets and a reconstruction of its philosophical and economic foundations.

After the meeting Julian Le Grand wrote to the then General Secretary of the Fabian Society, Ian Martin, to suggest that a group be set up to meet on a regular basis and to begin rethinking and reconstructing socialist ideas. There was an enthusiastic response to the idea, and the Socialist Philosophy Group was set up by the Fabian Society under the joint convenorship of Le Grand, Martin, and Plant.

At the first meeting of the Group, David Miller presented a paper on market socialism. This aroused considerable interest, and the topic formed the basis of many subsequent discussions in the Group. During the course of those discussions it became apparent that several of the Group's members had a common interest in the ideas grouped under the umbrella of market socialism—ideas that seemed to be worthy of further development and of dissemination to a wider audience. This book is the result.

Although each chapter in this book is individually authored, it is in a real sense a collaborative—dare one say, socialist—effort. Each contributor has read, and commented extensively on, the others' contributions. We have held several meetings to discuss the material and to iron out differences. We have not always been successful at the latter; but the differences that remain are small and do not, we believe, detract from the book's intellectual coherence.

We have tried to make the book accessible to as many people as possible. To this end we have tried so far as possible to avoid technical jargon and specialist analyses. We have also tried not to burden the reader with excessive footnotes and references. Footnotes have been kept to a minimum; the references are collected at the end of the book.

We have many debts. We owe a special one to the Executive and staff of the Fabian Society, particularly the two successive General Secretaries, Ian Martin and John Willman, for their encouragement and support of the Socialist Philosophy Group over the years. We have benefited greatly from all the contributions to the debates on the topic within the Group itself, including (and perhaps especially) the contributions from those who profoundly disagree with our arguments. Many other friends and colleagues have helped us develop and refine our views. Jane Dickson bore the brunt of the organization and additional typing. Finally, our families have had to tolerate the demands of the project, as well as those of our normal occupations and preoccupations. To them all, our deepest thanks.

PETER ABELL
SAUL ESTRIN
JULIAN LE GRAND
DAVID MILLER
RAYMOND PLANT
DAVID WINTER

Summer 1988

Contents

About the Authors

PETER ABELL is currently Head of Department of Sociology at the University of Surrey. He was formerly Director of Research at the Social and Economic Studies Unit at Imperial College, University of London, and Professor of Sociology at the University of Birmingham. His interests are in co-operatives, particularly in developing countries. He is the author of two books, *Small Scale Industrial Co-operatives in Developing Countries*, with N. Mahoney (Oxford University Press, 1988), and *Establishing Support Systems for Industrial Co-operatives: Case Studies from the Third World* (Gower, 1988).

SAUL ESTRIN was educated at the Universities of Cambridge and Sussex, and has worked at Southampton and Cornell Universities. Since 1984 he has been in the Department of Economics at the London School of Economics. His research has focused on comparative economic systems, with books on indicative planning and self-management in Yugoslavia, as well as numerous papers on producer co-operatives in Western Europe. He is currently working on privatization in Britain and France.

JULIAN LE GRAND is Professor of Public Policy at the School for Advanced Urban Studies at the University of Bristol and a Research Director at the Suntory Toyota International Centre for Economics and Related Disciplines at the London School of Economics, where he co-directs the Centre's Welfare State Programme. His books include *The Strategy of Equality* (Allen and Unwin, 1982), *The Economics of Social Problems*, with Ray Robinson (Macmillan, 1984), and *Not Only the Poor: The Middle Classes and the Welfare State*, with R. Goodin (Allen and Unwin, 1987). He was co-founder with Ian Martin and Raymond Plant of the Socialist Philosophy Group.

DAVID MILLER is Official Fellow in Social and Political Theory at Nuffield College, Oxford. Among his publications are *Social Justice* (Oxford University Press, 1976) and *Anarchism* (Dent, 1984). He has been working for some years on the ethical underpinnings of market socialism, and the results of this work will shortly be published in book form under the title *Market, State, and Community* (Oxford University Press).

RAYMOND PLANT is Professor of Politics at the University of Southampton. He is the author of books on Hegel, Political Philosophy and Welfare, Citizenship, Conservative Capitalism, etc., and the Fabian pamphlet *Equality, Markets and the State*. He was co-founder with Julian Le Grand and Ian Martin of the Socialist Philosophy Group.

DAVID WINTER is a Lecturer in the Economics Department of the University of Bristol. He has recently published papers on the consumption goods markets of centrally planned economies and in particular of Poland, and the determinants of British government expenditure. He is currently a Senior Research Fellow at the the Suntory Toyota International Centre for Economics and Related Disciplines at the London School of Economics, where he is a member of the Centre's Welfare State Programme.

Market Socialism

Saul Estrin and Julian Le Grand

THIS book has two aims. The first is to 'couple' markets and socialism. We hope to show that markets can be used to achieve socialist ends. The use of markets in this way is what we mean by market socialism, and the chapters in this book show how this may be done in a variety of contexts. An important corollary is the 'decoupling' of capitalism and markets. While it may be impossible to have capitalism without markets (at least in part because, as we argue later, all industrial systems, whether capitalist, socialist, or mixed, inevitably use markets of one kind or another), it is perfectly possible to have markets without capitalism.

The second aim is to start the radical reorientation of socialist thinking that is required by a proper understanding of market socialism. The market mechanism is the most efficient way of co-ordinating decentralized economic decision-making. This means eschewing the tendency to centralized intervention in the economy characteristic of socialist parties. Perhaps more fundamentally, it means a change in our understanding of the appropriate role of the state. Mistrust of the intentions of bureaucrats and the effectiveness of public interventions leads market socialists to seek to err on the side of *laissez-faire*. If one wishes to ensure socialist outcomes from a market mechanism, one must alter the environment in which markets operate to ensure that such outcomes are in the private interest of individuals, rather than use the state to impose the public interest from above. This leads some of the chapters that follow to focus on institutional changes and legal reforms conducive to the socialist vision.

One of the things that has made our task particularly difficult is that socialists have often been careless in distinguishing between ends and means. Socialism has a well-defined set of *ends*: for example, preventing exploitation of the weak by the powerful, greater equality of income, wealth, status, and power, and the satisfaction of basic needs. But many socialists have conflated these with a particular set of *means*, such as the state ownership of production or centralized planning of the allocation of resources—means that have become objectives in their own right. So socialism is identified with, for instance, the attainment of greater equality through planning or the elimination of exploitation through the nationalization of industry. But a key theme of this book is that there is no logical reason for these traditional identifications to hold. There is nothing intrinsic in planning that implies equality or in nationalization that eliminates exploitation. Nor, by extension, is there anything intrinsic in markets that prevents them from being used to achieve those ends.

In this introductory chapter we provide a summary and partial synthesis of the arguments developed in more detail in the rest of the book. Readers interested simply in obtaining a basic understanding of the principal ideas, together with some of their implications for policy and practice, should read this chapter and then Chapters 2, 7, and 8. Those who want to come to grips with some of the more complex economic and philosophical issues raised by the concept of market socialism should concentrate on Chapters 3, 4, 5, and 6.

A final point before we begin the discussion proper. This book is not concerned with providing short-term solutions to contemporary British economic and social problems. There will be virtually no discussion of current issues such as unemployment, inflation, poverty, the inner cities, racism, sexism, or the nuclear bomb, important as these are. Instead it is an attempt to take the argument away from the immediate concerns of the present, to the original issues of the British socialist debate—egalitarianism, an end to class exploitation, breaking the transmission of wealth inequalities, ownership arrangements, and the organization of the production process. We hope that, through the analysis of a decentralized economic mechanism such as the market, we can cast a fresh light on these fundamental problems and, in doing so, construct an intellectually coherent model of the socialist ideal.

THE MERITS OF MARKETS

We begin with the economic advantages of markets. Perhaps the greatest of these is that, when they work well, they are an excellent way of processing information, while simultaneously providing incentives to act upon it. In a competitive market, if a good is in short supply relative to demand, the price of that good will rise, indicating to producers that there are profits to be made if they produce more. Since market producers are generally motivated by profit, this is an incentive to which they will respond. If there is over-production of a good, then the price will fall, indicating to producers that they should switch their production to something else that is more in demand. There is no need for long communications between retail or wholesale outlets and central planners giving daily updates on the shortages or surpluses that are appearing; no need for detailed central planning directives to productive enterprises telling them how much or how little to produce.

For much the same reasons, markets tend to encourage innovation both in production techniques and in the goods themselves. Producers are constantly on the look out for ways to increase their profits by stealing a march on their competitors; creating a successful new product or introducing a cheaper production process are obvious ways in which this can be done.

Competitive markets also have the advantage that they disperse economic power. People have a range of other people with whom they can deal; they are not at the mercy of an awkward manager or a recalcitrant clerk. If they do not like the service offered by a particular supplier, they can go to another providing a better quality service. Moreover, the very process of switching from unhelpful to helpful suppliers will encourage the survival of the latter at the expense of the former.

David Miller in Chapter 2 notes these efficiency advantages, but goes on to appeal to other key values in the defence of markets: freedom and democracy. Markets, he argues, promote freedom in (at least) three ways. First, and most obvious, the dispersal of economic power means that people have greater freedom of choice over what and where to buy. Second, they also have a much greater freedom over when and where to work. Obviously, this freedom is limited by the availability of work (something that may be greater under planned systems), but, within that context, the use of financial incentives is likely to be more freedom-promoting than the labour direction that is an unavoidable feature of planned economies. Third, markets promote freedom of expression. Without an independent source of economic power, people attempting to propagate views that run contrary to those of the state will only be able to do so by the use of the state's resources—and he who pays the piper calls the tune.

Of course, markets also have their failures. Many market activities impose costs on people other than the immediate participants: environmental pollution is an obvious example. Conversely there are activities that confer benefits on non-participants and will tend to be under-provided if left to the market: immunization is an example. An extreme form of the latter phenomenon are so-called 'public goods' like defence or law and order. Other problems arise from technological factors, such as economies of scale; a competitive market in an industry with economies of scale will, as soon as one firm gets an edge on the others and begins to grow, rapidly degenerate into a monopoly. Even more important, in areas crucial to people's welfare such as education and health care, suppliers of a service (doctors, teachers) are often much better informed than the people they serve; hence users of these services cannot properly assess quality and are therefore open to exploitation.

While market failures such as these show the need for appropriate intervention, Miller responds more critically to three other alleged disadvantages of markets: they respond, not to real 'needs', but to superficial demands, often created by market producers themselves; they encourage anti-social, selfish behaviour; and—perhaps for socialists the most crucial of all— they create a morally arbitrary, and therefore unjust distribution of income. With respect to the first, he points out that in non- subsistence economies the notion of 'need' is problematic: who is to decide what people need, the planner, the social scientist, or the people themselves? Nor is the fact that some wants are market-created conclusive. Every economy operating at above subsistence level has to cater for wants that are created by the society itself; there seems little reason to regard the wants generated in a market economy as any more psychologically suspect than those generated in a feudal, tribal, or even centrally planned society.

With respect to the selfishness of the market, he accepts that markets are unlikely to be compatible with a 'monolithic' sense of community—one in which all relationships are of the same self-denying character, subordinating individual interests for the good of the community. However, he does not find such a community, with its denial of individual personality and life- style, attractive. Moreover, as the existence of villages and indeed market towns indicates, the existence of markets is perfectly compatible with looser communities—ones that allow their members a variety of interactions with each other, includ- ing love, friendship, and compassion, and exchange relationships including market ones.

Finally, the injustice of the market. Miller points out that market distributions are not necessarily always unjust; markets, for example, will, other things being equal, reward hard work and thrift, outcomes that accord with some (desert) notions of fairness or justice. However, it has to be acknowledged that they will also reward luck: the luck of being born into the right family, the luck of owning a house near the new underground extension. Moreover, the inequalities thus generated tend to be cumulative over time, with the owners of large fortunes having the economic power to defend and to enhance their privileges.

But many of these failings can be put down in large part to market *capitalism* rather than to markets themselves. A world where there are few owners of capital, and capital owns and controls enterprises, is a world of continuing inequality and exploitation. Since there is no intrinsic link between capitalism and markets, this raises the question as to the alternatives. Is it possible to devise a market system that can attain socialist ends and even incorporate a particular version of socialist means? In other words, is it possible to create market socialism?

MARKET SOCIALISM: A CONTRADICTION IN TERMS?

There are those from both Left and Right who, when faced with the question at the end of the last section, would answer with a firm no. For them markets and socialism are at the opposite ends of the political and economic spectrum; their conjunction in the phrase market socialism is a nonsense, a contradiction in terms. Other chapters in this book are aimed at disproving this assertion by illustrating in practice how market mechanisms can be used to attain socialist ends. However, there is a fundamental challenge, emanating particularly from the New Right, to the philosophical basis for market socialism that has to be addressed before we can proceed further. This is taken up by Raymond Plant in Chapter 3.

Socialism is conventionally identified with 'end states' or outcomes. That is, a socialist society is one where social outcomes are specified according to a particular model—one based on fundamental socialist values such as justice, (positive) freedom, or community. Now liberal defenders of the market have argued that markets cannot be used to attain socialist outcomes. This is because, although markets consist of human actions, they do not produce outcomes that are of human design. The distribution of income, or the pattern of consumption, that emerges from the operations of markets is unintended, undesigned, and unforeseen. To complain that because some people end up with more and others with less this is unjust, curtails positive freedom, or is anti-communitarian is like complaining about the injustice or illiberality or anti-socialness of the weather: a literal 'nonsense'. The only 'end' that markets can attain is essentially a negative one: negative liberty or the absence of intentional coercion.

The use of markets to attain socialist ends has another set of problems associated with it. The terms in which those ends are expressed—terms such as 'needs' or 'social justice'—have a wide variety of interpretations, each of which, according to the New Right, is equally defensible or indefensible. Socialism, New Right thinkers argue, involves the imposition of one particular interpretation of each of these terms on everyone—a procedure that has no moral justification and, moreover, is unlikely to be practically sustainable without the use of unacceptable levels of force or coercion. Markets, by way of contrast, allow people's preferences (including their own interpretations of moral terms) free rein. Markets and socialism thus seem quite incompatible.

To this there are a number of responses. First, the very fact that the New Right thinkers justify the operation of markets with respect to one particular end—negative liberty—suggests that in fact even they believe that market outcomes are not morally neutral. Second, we know that markets can be regulated, supplemented, or even supplanted entirely by government action. Hence any decision to allow them to operate freely involves an acceptance of the morality of the outcome. Thus to judge the outcomes of market operations according to any particular set of values is not a 'nonsense'.

Plant provides a third argument. Market socialism denies the link between socialism and outcomes. Indeed, in this respect, market socialists accept a large fraction of the liberal case: that people on the whole should be left to determine their own idea of the 'the good' and indeed of the 'good life'. What market socialism does require—the aim of market socialism—is greater equality at the *beginning*: that people enter markets on an equal footing. They are committed to equality at the starting-gate, not equality at the end. And why should there be equality at the starting-gate? This is justified by the very arbitrariness of moral theorizing to which the liberal theorists themselves draw attention; for, in the absence of any well-defined or consensual theory of merit and desert, there is no justification for not distributing resources equally.

Does this mean that market socialists have no views about outcomes—or, more correctly, that they actually accept all market outcomes as in accord with their values—so long as they have been attained from equality at the starting-gate and so long as the market operations were themselves fair? A radical interpretation of market socialism is to answer yes; if one of the motivations behind the market socialist enterprise in the first place is to give people greater positive freedom, then we cannot object to the outcomes, even if they are ones that we personally do not like. Another view would be somewhat more eclectic: to allow markets to operate, but to use regulation as appropriate to achieve certain well-specified outcomes (e.g. permit education vouchers while at the same time enforcing a national curriculum). But this raises wider issues concerning the overall objectives of a socialist system and how best to achieve them—issues addressed in Chapter 4 by Peter Abell.

ENDS AND MEANS

Abell begins by discussing the appropriate ends to which socialists might subscribe. Inevitably, these must include equality in some form; all socialists must be egalitarians, even if not all egalitarians are socialist. However, like some other contemporary socialist thinkers (see, for instance, the recent works by Brian Gould, 1985, and Roy Hattersley, 1987), Abell sees no reason why the Left should let the values of freedom and efficiency by appropriated by the Right. Rather, socialist principles should be formulated that take account of equality, freedom, and efficiency.

He begins his search for such principles with those that have traditionally guided socialist thinkers: from each according to ability, to each, initially according to ability (in the early stages of socialism, while acquisitive values still prevail) and then according to need (when acquisitive values have diminished or disappeared). Abell points out that neither principle takes any account of liberty; and that the second is silent on the question of incentives for efficiency, while the first assumes a particular structure of incentives that may bear little relationship to reality.

Instead he proposes an alternative: the equalization of positive freedoms in production.

This technical phrase may be roughly interpreted as implying that, so far as possible, everyone should have equal resources or productive capacities—an idea that has similarities to the arguments of Gould (1985) and Le Grand (1984). If, for instance, an individual was born with less ability, physical or intellectual, than the average, he or she should be 'compensated' with other resources such as education or material assets. In contrast with most Western social systems, this would imply that more education should be given to the less rather than the more able, and that inheritance of wealth should be directed at those disadvantaged in other respects rather than at those already privileged.

Abell acknowledges that policies designed to equalize productive capacities in this way may have short-term disincentive effects and thus impede economic efficiency. Hence he recommends that the policies concerned be introduced gradually, allowing a progressive acceptance of the redistributive effects.

But what would these policies be? Abell does not discuss these in detail, for they are treated elsewhere in the book. However, he argues that socialists ought to be ethically neutral between equal access to productive assets which are socially owned and those owned privately but equally distributed. The choice between them is primarily a matter of efficiency. On this basis, Abell does not support nationalization of the type sanctioned by old-style interpretations of the celebrated Clause 4 in the Labour Party constitution; he is also sceptical of the value of more recent ideas of economic restructuring such as the 'share economy' proposed by Martin Weitzman (1984). Instead, he advocates resource-equalization measures of the kind mentioned above, coupled with competitive producers, each organized according to one of several possible forms of producer democracy. Here he favours the labour–capital partnership, where both labour and capital possess share certificates which entitle the shareholder to a dividend, and where the enterprise concerned is controlled by directors elected in equal numbers by labour and capital. This would, in his view, go a long way towards the establishment of the ultimate socialist goal: an equitable, free and efficient society.

We have seen that markets have many attractive properties and, in particular, that markets can actively promote rather than hinder socialist objectives provided that the distribution of income, wealth, and opportunities are right. But what of the traditional socialist argument that markets tend to aggravate the evils of capitalism, and therefore need to be overridden by a central planning mechanism? This is the subject of Chapter 5 by Saul Estrin and David Winter.

The chapter has two objectives. The first is to highlight the fact that market socialists are not unaware of the systematic deficiencies of the market mechanism from the socialist point of view. Miller has already pointed out the informational and incentive strengths of the market mechanism. The point Estrin and Winter stress is that, provided the market system is competitive, production for profit can be *socially* as well as privately efficient. The traditional socialist distinction between production for profit and social production has no basis in economic theory for the vast majority of goods on the market.

Exceptions to this rule are termed market failures, and market socialists would have a longer list of examples than apologists for the New Right. For example, allocation by markets can fail completely when the good under consideration is consumed collectively, or has to be produced monopolistically to exploit economies of scale. Serious allocative problems will also emerge if there is complete *laissez-faire* in the allocation of capital or goods with major spillover effects. The other side of the decentralization coin is the 'anarchy of the market', with excessive price volatility and waste from duplication of effort and capacity. The informational and incentive advantages of the market system may be offset by systematic overshooting of prices and inadequacy of response to purely material motivations. Markets may therefore be weak in inducing non-marginal changes in the structure of the economy. Finally, markets have a tendency to aggravate wealth and income inequalities, and the problem cannot be dealt with entirely by taxation, at least in the short term, because of the effects on incentives.

This daunting list does not tempt Estrin and Winter to abandon the market as the principal allocation mechanism. In the end, these disadvantages are not sufficient to outweigh the gains in efficiency from decentralization. However, market socialists will want to prevent as many as possible of these problems from emerging, by altering property rights and by adjusting the legal system so as to provide incentives for people to choose more desirable patterns of behaviour. In the limit, market socialists may even have to override certain free market outcomes. Preferably this will be in a decentralized way by sponsoring the emergence of non-market institutions to deal with specific problems, but, if necessary, direct government intervention will be used. 'Indicative planning' can be used to stimulate such private initiatives. This involves a decentralized and democratic process of consultation and discussion to devise a guide to medium-term economic development in the medium term. Creating the institutions of an indicative planning process could be an important example of institutional intervention by a market socialist government.

Indicative planning acts as a complement to, rather than a substitute for, the market mechanism. In contrast to central planning, it has no implementation phase. It allows individuals to make their economic decisions in the knowledge of what their suppliers, buyers, and competitors will do. More importantly, it allows a market socialist government to co-ordinate its range of policy interventions to ensure market outcomes.

Estrin's and Winter's second objective is to argue that the case for market socialism rests on more than the attractions of markets. It is also based on the failures of the principal alternative: central planning. Their discussion examines planning both from a theoretical perspective, and with reference to Soviet and Eastern European experience. It is relatively easy to establish that central planning of an entire economy is unfeasible. Planners do not have enough information to construct plans which are internally consistent, and the other actors in the economy—workers, consumers, and particularly enterprise managers—have no incentive either to provide the correct information or to implement the plans properly. Estrin and Winter also point to the dangers of totalitarianism inherent in the overcentralization of a planning mechanism.

The experience of central planning in the Soviet Union and Eastern Europe as a means of attaining socialist ends does not inspire confidence in the ability of such planning to eliminate waste or encourage efficiency. Planners attempt to resolve the informational and incentive problems by 'tautness'—setting plans that are at or beyond the limits of the enterprises' capacity. This creates shortages throughout the economy, with internal inconsistencies resulting from the failure of firms to meet their targets. The economy only functions at all in such a system because of the emergence of black and grey markets. Socialists attack the 'anarchy' of markets, but there is also an anarchy of central planning, with a perpetual sellers' market, speculation and corruption in black markets, and extensive waste and poor quality outside priority areas.

Despite the failure of central planning to sustain a reasonable growth in living standards over recent years, attempts at reform in, for example, Poland have tended to make matters even worse so far. The entrenched bureaucracy at the heart of the planning system is a highly conservative élite, as unwilling to forgo its privileges as any of its capitalist counterparts. It is small wonder that centrally planned economies have done little to eliminate inequalities in wealth and privilege; indeed, in some respects these inequalities are greater than in the West.

So, even though markets have severe deficiencies, Estrin and Winter believe that central planning does not offer a viable alternative. We are forced back to the market mechanism. The argument here is quite subtle. In practice, all economies use markets and all use planning, to a greater or lesser degree. So-called market economies plan various specific fields of activity—within the welfare system, education, or multi-divisional firms. Similarly markets emerge in planned economies, whether legal or illegal, to fill allocative gaps. The question is which of the two mechanisms is to be the *principal* method of allocating resources: market or plan. Estrin and Winter argue that, if markets are used as the principal economic mechanism, planning can be used as and where it is necessary. If planning is used, the market mechanism is debilitated, and is too weak to pick up the pieces. Central planning institutions necessarily suppress and damage the market mechanism and the key characteristics on which it thrives—risk-taking, entrepreneurship, and competitiveness. It is better to make markets the principal exchange mechanism, supplemented by non-market mechanisms should the need arise. When these break down, a well-developed market system will be available on which to fall back.

SOCIALIST ENDS AND CAPITALISM

If socialists cannot reject markets, perhaps they should instead embrace capitalism in its entirety. Socialist worries about inequality could perhaps be assuaged by taxes and subsidies, with the welfare state ensuring minimum standards in education, health, and living standards. Social corporatism and the enhanced power of unions would circumvent problems of capitalist domination at the workplace. This vision has been inspired by the successes of social democracy in Sweden and Austria, and has helped to motivate much of the British Labour Party's thinking since the Second World War (see in particular Crosland, 1964). It is challenged and decisively rejected by David Winter in Chapter 6.

Winter seeks to expound the fundamental objections that socialists have to the capitalist system, and thereby to indicate the form that changes will have to take if one seeks to introduce market socialism. The analysis takes him back to the problems that left-wing thinkers have traditionally identified with the capitalist system: exploitation, the dynamic process which continually regenerates inequalities of income and wealth, and the domination of labour by capital in the workplace. He finds little merit in the traditional socialist solution to these problems—nationalization. Rather he points to fundamental changes in the property-right system which will impede or eliminate these exploitative processes despite the operation of markets.

Exploitation of workers is traditionally thought to arise because capitalists own the means of production, initially by direct possession but most commonly now through limited liability companies. The Marxist notion of exploitation rests on the labour theory of value, which shows that, when workers sell their labour services to capitalists, its value to the capitalist is greater than the value of goods the workers can purchase with their wages. This view has recently been challenged by John Roemer (1982), who establishes that this concept of exploitation becomes harder to sustain as social heterogeneity increases. In particular, he provides a theoretical counter-example to the Marxist view that capitalists always exploit workers by noting the possibility of rich workers exploiting poor capitalists. The straightforward link between inequalities in wealth and exploitation can break down when people differ with regard to inherited abilities and tastes.

Winter uses this rather abstract argument to distil the essential characteristics of capitalist exploitation from the ancient theology of the labour theory of value. Exploitation in a capitalist society arises from two sources: the differential ownership of productive assets, and the dispersion of skills and talents across the population. Even if all income differences are not entirely eradicated, socialists would still wish to eliminate the inequality that arises from differences in the ownership of the means of production, both because these are the principal source of inegalitarianism, particularly over time, and because they underlie the domination of workers by capitalists in the workplace.

Socialists have usually sought to reduce this problem by the piecemeal nationalization of industry. Winter argues that the experience of public ownership in most Western economies has not been a happy one. It is not clear which assets should be nationalized, nor what should be done with them once they are in public hands. The public sector of most Western countries is therefore typically a ragbag of utilities, public service corporations, defence contractors, oil companies, previously bankrupt capitalist firms, and miscellaneous others united only by their ownership structure. Since the purpose is to prevent others from owning the assets, rather than to achieve anything with them itself, the state often runs these firms badly, so problems of inefficiency and bureaucracy generally emerge. Political control produces powerful interest groups but rarely enhances economic efficiency, nor provides distinct benefits to the firm's workers.

If nationalization is not the panacea, can the exploitation problem be resolved by 'egalitarian capitalism'? Put another way, is there something inherent in the capitalist system which generates an unequal distribution of capital, and therefore exploitation? Winter answers this question in the positive. Capitalism has traditionally been characterized by the scarcity of capital and the abundance of labour, explaining the relatively high returns paid to the former. As accumulation reduces the return to capital, new scarcities are generated by technical advance. Moreover, the uncertainties surrounding the generation of profit ensure that the surviving owners of capital are relatively rich; the unsuccessful capitalists, who makes losses rather than profits, have their assets scrapped and revert to being workers. Successful capitalists on the other hand can diversify their portfolios to spread their risks, which further increases the concentration of ownership. These persistent relative scarcities of labour and capital mean that workers will, on the whole, earn a relatively small share of the cake, and that capitalists will have the power to arrange and control the lives of workers at the workplace.

Having isolated the problem, Winter argues that the only way of reforming a capitalist economy is by changing the legal framework that supports it. Wholesale expropriation via nationalization has limited benefit. He has sympathy for the 'poll grants', discussed in Chapter 8 by Julian Le Grand, but feels the long-run effects will be slight given the inherent dynamics of the capitalist system.

Winter also notes that capitalist acts can bring significant social benefits—for example, the widespread diffusion of new production processes of products. As he points out, with appropriate limits it is not clear that one would want to rule out capitalist acts between consenting adults altogether. So firms can be capitalist up to a certain size, but the bulk of ownership cannot be private in a socialist society.

In practice, most large firms are not privately owned, but owned by their shareholders. Winter sees this separation of ownership and control as an important source of inefficiency in contemporary capitalism. But it does open the possibility of altering ownership arrangements without the efficiency losses that could result from the appropriation of the assets of 'heroic capitalists'. Changes in the Companies Act to eliminate the limited liability company are relevant here. Winter concludes that capitalist or managerial domination in the workplace is best resolved by a system of workers' control: the subject of Chapter 7.

WORKERS' CO-OPERATIVES

Workers' co-operatives are intellectually fashionable across much of the political spectrum and deservedly so. In Chapter 7 Saul Estrin spells out their advantages; but he also points to their limitations, and suggests ways in which these may be overcome.

The first merit of co-operatives is that they eliminate the exploitation of labour by capital. Rather than the owners of capital hiring labour, labour hires capital. The means of production become a tool of labour instead of its master.

Second, co-operatives are democratic. In contemporary Western economies, there is a sharp contrast between democracy in the political process and autocracy in the workplace. The latter creates dissatisfaction and alienation, leading to, on the individual level, obstructive behaviour on the factory floor, absenteeism, shirking, and high labour turnover, and, in a unionized environment, industrial militancy. By contrast, in a co-operative, power is spread throughout the enterprise with each member formally having equal voting rights, and a chance to share in managerial functions. Hours and other aspects of working conditions can be altered in line with workers' desires. All this can create a highly positive attitude to work and an increased commitment to the activities of the enterprise.

This leads to the third advantage of co-operatives: their potential for increasing efficiency. This arises in part because of the reduction in alienation and the accompanying increase in work commitment just described. The enterprise may also benefit from being better placed to draw on the expertise of the shop-floor, a valuable resource generally ignored in the hierarchical capitalist firms. But the increase in work effort and thereby efficiency may also come about because each member has a greater stake in the profit of the enterprise. They are not working simply to provide for shareholders' dividends: the rewards from the extra hour worked at the end of the day go to those who work it.

The potential of co-operatives for increased efficiency has been rather obscured in Britain by the spectacular failure of the Meriden, Scottish Daily News, and Kirkby co-operatives sponsored by the Labour Government in the 1970s. However, these seem to have been unrepresentative. Italy, France, and Spain have large and successful co-operative sectors. Even in the United Kingdom, Estrin points out, the number of co-operatives has increased from less than twenty in 1975 to perhaps, 1,600 today; moreover, their failure rate seems to be lower than that of other types of small businesses.

A fourth advantage of co-operatives is that they are likely to be more egalitarian than their capitalist counterparts. The distribution of profits among the work-force is likely to be more dispersed than among shareholders; and, although some pay differentials remain within the work-force, Italian, French, and Spanish experience suggests that they will in general be much smaller than in the conventional capitalist firm.

But, as Estrin goes on, neither the undoubted merits nor their currently fashionable status should blind us to the defects of co-operatives. First, their concern for workers' welfare means that they tend to adjust output less than capitalist firms in response to changes in demand and in cost conditions. In good times, co-operatives will not adapt sufficiently to high demand or technological changes. In bad times, co-operatives are ill-suited to the hard decisions involved in fundamental capital and labour restructuring.

This would not be so much of a problem if it were relatively easy to start or close co-operatives, for then the forces of competition would weed out the sluggish co-operatives and the overall efficiency of the economy be maintained. But, Estrin argues, in practice it is not easy to form or close co-operatives. They are difficult to form because their potential members have to find each other. Moreover, entrepreneurs who have found a profitable opportunity are likely to want to exploit it themselves, not share it with their work-force. They are difficult to close because, due to the very commitment they engender, their work-force is often prepared to accept sacrifices that permit enterprises to operate at levels well below what would be economically viable in a more conventional context.

A second major problem concerns the tendency of co-operatives to underinvest. This arises because, in co-operatives collectively owned by the work-force, capital investment has to compete with pay for the distribution of profits. If workers cannot take their 'stake' out of the co-operative when they leave, then the only incentive they have for investing in the enterprise is the extra earnings they receive. That is, they benefit only from the return on the investment; the principal becomes part of the collective assets of the enterprise. Necessarily in this situation they will invest less than if the investment were being undertaken by conventional capitalists, since the latter will not only benefit from any return but can recoup the principal by selling their shares or if necessary the enterprise itself.

Further, if workers invest their savings in the enterprise by forgoing pay increases out of profits, they are obviously less vulnerable to the power of capital than in traditional capitalist firms. But they are, if anything, more vulnerable to the power of the consumer; for capitalist firms have shareholders who absorb some of the risk and can cushion workers from short-term fluctuations in market demand. In a fully co-operative economy, a fall in the demand for an enterprise's product could result in workers losing both their livelihood and their savings. Again, as well as being socially undesirable, this is likely to discourage investment.

These difficulties—unresponsiveness and underinvestment—seem pretty damning. However, Estrin argues, they are not endemic to *all* forms of co-operatives, but merely to certain types. One solution to the underinvestment problem, for instance, is to adopt a model, more prevalent in the United

States than in Europe, where the workers are given equity shares in the enterprise. Another model, one that addresses both sets of problems, is the labour–capital partnership, discussed briefly by Abell in Chapter 4. Here owners of capital take out equity shares in enterprises; but both labour and capital are represented on the board, with neither side necessarily holding a majority of voting rights. The fact that capital is represented on the board means that conventional notions of economic return and viability will play a more important role in determining the enterprise's activities, thus increasing its responsiveness to market conditions and its investment potential.

Estrin, however, prefers a more radical solution. He proposes the establishment of competing holding companies in which the ownership of productive capital would be vested. These would lend capital to enterprises at the market rate of interest. The holding companies would be empowered to set up new co-operatives in areas with profit potential; they would also be able to close down struggling co-operatives if their wages fell below a prescribed minimum. The holding companies themselves would be owned by equity shareholders, by the state, or by other co-operatives.

This system should overcome most of the difficulties faced by a co-operative economy. But there remains the problem of the transition. How do we get there from here? An interesting suggestion, originally made by Winter and developed by Estrin, is to use the structure of existing publicly quoted firms. The existing ownership arrangements would remain unchanged; but the head office of each firm would be converted into the holding company, while the firm's plants would be transformed into self-managing co-operatives and given their 'independence'. They could choose to remain with their original head office, now a holding company, paying to that company the market interest rate on the value of their productive capital; alternatively, if they preferred, they could shift to another head office or holding company if that offered them a better deal.

MARKETS, WELFARE, AND EQUALITY

In the final chapter Julian Le Grand considers two topics, both concerned with the relationship between market socialism and welfare. First, he discusses the potential for markets in what is, outside the family, perhaps the largest area of non-market activity in most Western economies: the provision of welfare services, such as education, health care, housing, social care, and social security. In recent years the welfare state has come under an unprecedented barrage of criticism, being accused of inefficiency in its use of resources; of being unresponsive to the needs and wants of its users and more concerned with the interests of its employees; of creating dependency and undermining economic and other incentives; and of failing to achieve equality, both within specific welfare areas such as education and health and within the wider society.

Many of these problems have been wildly exaggerated; yet, as Le Grand argues, the fact that things are not as bad as they are often made out to be does not mean that all is well. Undoubtedly, much of the welfare state is inefficient, unresponsive, and inegalitarian. Now the conventional wisdom concerning markets is that they tend to encourage efficiency and responsiveness, but exacerbate inequality. But is this always true? Is it possible to introduce market-type welfare reforms that promote efficiency and responsiveness but do not increase—and perhaps even reduce—inequality? It is to these questions that this final chapter is addressed.

Le Grand begins by looking at the case for the most extreme 'reform' of the welfare state: its replacement in all areas of welfare by the private market unimpeded by any form of government intervention. He discusses some well-known problems with the use of unrestricted markets for the provision of welfare services, including the existence of 'external' benefits associated with many of these services, the imbalance of information between suppliers of services and their clients, and the fact that the distribution of services would be determined primarily by the distribution of market incomes. A further problem to which he draws attention is the possibility of family exploitation; it is arguable that a major purpose of the welfare state is to protect individuals from their families.

So full-scale privatization is not the answer. But what of reforms that incorporate elements of markets, but stop short of complete replacement of the welfare state? Le Grand discusses two of these in some detail: vouchers, and tax-related charges. Although versions of these ideas have been colonized by the Right, he points out that some of them also have real merits in socialist terms, especially if they are adapted appropriately. A voucher system, for example, that discriminated in favour of the poor would increase their power relative to that of welfare providers as well as relative to other, richer welfare users—an outcome that seems quite consistent with socialist ideals. Similarly, a charge for services that was incorporated into the tax system—a 'user tax'—could promote both efficiency and fairness—again, a socialist outcome.

However, there are serious difficulties also associated with the application of the various ideas in the welfare area, and Le Grand warns that it is important not to adopt them wholesale. Instead, he recommends that, where possible, some limited experiments are undertaken so as to assess in practice the extent of the problems and whether they outweigh the advantages.

The second major topic in the chapter concerns the potential impact of the tax and welfare systems on wider economic and social inequalities. In particular, Le Grand focuses on the redistribution of wealth, a topic that has been rather neglected in recent years. He advocates the introduction of a lifetime capital receipts tax, the revenues from which are used to finance a 'poll grant': a capital grant to everyone on reaching the age of majority. This would go some way towards the attainment of equality at the starting-gate—an aim which, as pointed out in Plant's and Abell's chapters, is an essential precondition for market socialism.

WHAT MARKET SOCIALISM IS NOT

After an explanation of what market socialism is, it seems important to give some discussion of the things it is not. One of the most important of these concerns racial and gender exploitation. Market socialism is race- and gender-blind. Individuals are consumers, savers, workers—not men or women, black or white, each with special privileges, problems, or interests. For some, blindness of this kind is the essence of a non-discriminatory society and hence will be one of market socialism's principal virtues; for others, it would be one of its gravest deficiencies.

However, even for the latter group, the 'deficiency' is one of omission rather than commission; the very neutrality of market socialism means that there is nothing within it that is contradictory to, for instance, equal opportunities or positive discrimination policies. Nor is there any reason why a market socialist economy should not operate effectively in the presence of an active enforcement of such policies. As Estrin and Winter stress, it is only the principal exchange mechanism which has to be the market. Things that are best left alone should be left alone. But if certain outcomes lead to public disquiet, market socialists have the full panoply of fiscal and legislative tools at their disposal to deal with them.

But, for its critics, perhaps the principal thing that market socialism is not is socialist. Some might say that the essence of socialism is the renunciation of competitive behaviour in favour of co-operation. Others would go further and argue that market socialism is antithetical to a socialist vision of the 'good life', where people behave in non-competitive ways, where the only things produced are 'socially necessary' and there is no vulgar consumerism.

The first of these criticisms—that market socialism discourages co-operation—is in part misdirected. Although a market socialist economy will have enterprises competing with one another, the enterprises themselves would be co-operative in some form or another; and to that extent co-operative behaviour would be encouraged.

The second has more force. It is true that, under market socialism, there is no overall, centrally imposed vision of the good life. Rather, each individual is free to work out his or her own vision. But this could be construed as an advantage rather than a disadvantage. Precisely what constitutes the good life is notoriously difficult to decide, particularly for people other than oneself. At the end of the day, the ultimate authority on what constitutes the good life has to be the person who is going to live it; and, under market socialism, that is where that authority is vested.

CONCLUSION

Capitalism places economic power in the hands of capital and its owners. Traditional socialism gives power exclusively to labour: the dictatorship of the proletariat, preferably exercised through a centralized authority. And the 'New Right'—actually better characterized as traditional liberalism—claims to locate power in the hands of the individual—particularly, the individual citizen and consumer.

Since we are all citizens and consumers, since most of us are (or have been, or will be) workers, and since the majority own, or would like to own, capital in some form (a house, savings accounts, pension rights, insurance policies, stocks and shares), it is not surprising that none of these traditional 'models' of how the economy should be organized finds universal favour. Full-blooded capitalism is unattractive because it exploits labour through its monopoly of employment and because it exploits consumers through monopolizing goods markets. Traditional socialism expropriates capital and subordinates the interests of consumers to the interests of the workers. Indeed, with its penchant for centralization, it is far from clear that even the interests of workers are properly taken care of. Liberalism puts people's livelihoods and their savings at the mercy of consumer taste and fashion; its emphasis on the narrow rights of individuals jeopardizes the collective activities of the community and hence the community itself.

What is needed is a model of society where power is more evenly distributed between these groups; where the interests of owners of capital, of workers, and of consumers are all taken into account with none taking automatic priority. It is the view of the authors of this book that market socialism comes closer to that ideal than any of the more traditional views.

We have not attempted to answer all the questions that market socialism raises, nor to respond to all the possible criticisms. Nor have we tried to present a complete description of Utopia. However, we believe we have offered a blueprint for a society that could be simultaneously egalitarian, non-exploitative, efficient, and free—that is, for a socialist society.

2

Why Markets?

David Miller

CURRENT attitudes on the Left towards markets as a form of economic organization range from the lukewarm to the positively hostile. Those who are lukewarm concede that markets may be unavoidable as a way of regulating the production and distribution of some goods and services, but they reserve all their enthusiasm for other issues: the redistribution of wealth, the public provision of essential services, and so forth. Others find no place for markets at all in their vision of the good society, seeing the socialist project as one of overcoming the market economy to the extent to which economic development and human psychology allow.

At one level this is hardly surprising. Faced with political opponents—the New Right ideologues whose theories are called in to support the policies of Mrs Thatcher and President Reagan—trumpeting the virtues of markets from the rooftops, it is understandable that the Right/Left and Market/Anti-market polarities should become identified with each other. But in fact the New Right position depends on a sleight of hand. Markets as a general way of organizing economic activity are equated with capitalism. Now it is certainly true that capitalism relies on markets, but what is distinctive about it is that the ownership of productive assets is concentrated in the hands of a few, with most people being hired as employees for a wage. It is quite possible to be for markets and against capitalism, and the Left has only itself to blame if it allows this possibility to be closed off by conventional usage. All too often it is: so, for instance,

For a more detailed discussion of many of the arguments in this chapter, see Miller (forthcoming). For alternative treatments of some of the issues, see Buchanan (1985) and Selucky (1979).

when a reporter looks at co-operative food shops operating on a market basis in Moscow, the first question he or she asks is: how can this be socialism? To which the equally simple answer should be: why not, if the co-operatives are democratically run and their assets socially owned?

There is, however, a deeper reason for the hostility towards markets still so often encountered on the Left, and to understand it we need to look briefly at the origins of the socialist tradition in the early nineteenth century. The early socialists reacted against the exploitation and impoverishment suffered by the newly formed working class at the hands of their employers, and also against the social fragmentation that resulted from the breakup of the pre-industrial village communities. In developing their visions of an alternative society, these socialists tended to emphasize, on the one hand, material equality and an increased standard of living for the labouring class, and, on the other, social harmony and co-operation in place of the conflict and competition of a capitalist economy. The social order envisaged was based on small local communities within which co-operative relations would prevail; Robert Owen's 'Villages of Co-operation' and Charles Fourier's 'Phalansteries' are prime examples. We see, then, that in this early form of socialism an attempt was made to combine the material benefits of industrialization with the social and human benefits of the pre-industrial communities. This clearly represented a potential source of tension within the various models put forward—a tension that was concealed in part because the early socialists made little attempt to analyse the economics of their proposed systems.

Marx, who inherited this tradition, distinguished himself from it by contrasting the 'Utopian socialism' of his predecessors with his own 'scientific socialism'. If, however, we ask precisely how Marx's theory differed from that of the 'Utopians', we find that the contrast lies in two main points. First, Marx's view of the transition to socialism was grounded in the material interests of the working class, whom he saw as suffering increasingly intense exploitation as capitalism lurched from crisis to crisis, whereas his predecessors had relied on the ethical appeal of their visions of socialism to all enlightened persons, including enlightened members of the upper classes. Second, Marx's theory was embedded in an ambitious account of historical development, inherited from Hegel, according to which the human species realizes its full potential only through an ordered series of stages, each of which develops in response to the inadequacies of its predecessor. This perspective allowed Marx to recognize the enormous and irreversible transformation of human relationships that capitalism had brought about. But when it came to describing the stages of socialism that would follow capitalism, Marx narrowed his focus to concentrate on the material benefits that would be handed down. His vision of communism—the higher stage of socialism—embodied the material achievements of capitalism, but little else. Indeed it had a great deal in common with those of his Utopian predecessors. It was expressed using terms such as 'alienation' that reflected Marx's background in German idealism; the substance, however, was familiar. The vision was of small-scale units in which conflicts of interest had been overcome, in which competition and profit-seeking had been replaced by co-operation, and in which the inequalities of capitalism had been superseded by the distributive maxim 'From each according to ability to each according to need'. Moreover Marx, though a far greater economist than any of his socialist predecessors, was equally silent about the economics of socialism itself.

Nineteenth-century socialism, to sum up, was a morally inspired vision of a society which negated the offensive features of capitalism, a vision drawing to some extent on the pre-industrial communities which capitalism had eroded. The quality of human relationships was of central concern, and this expressed itself in a fierce hostility to profit-making, always regarded by mainstream socialists as exploitative. From this perspective, the distinction between capitalism and other forms of market economy paled into insignificance. The emergence of genuinely communal relationships would spell the end of economic exchange.

This ideal has never entirely lost its hold, even on those whose practical ideas about the future of socialism have taken a very different path. As a vision, it is in many respects uplifting, and accounts for a good deal of the emotional pull of socialist ideology. But, for all that, it is deeply flawed. It tries to graft a form of community that is necessarily limited to pre-industrial economies on to the very different economic and social relationships that an industrial economy requires. And as a guide to the practical construction of a socialist society, it is of very little help, for socialism so understood amounts to little more than a series of negatives: 'no exploitation', 'no competition', 'no hierarchy', etc.

This practical hiatus paved the way for the later identification, by both friends and enemies, of socialism with state planning. For the early socialists, the state was at best a means of transition to a form of society which was often itself conceived as stateless (for instance by Marx). But, in the absence of any detailed guidance, it was all too easy for later generations to assume that the degree to which a society was socialist could be measured by the extent of state involvement in the production of goods and services—by the number of industries taken into state ownership, the percentage of GNP devoted to the welfare state, and so on. Because socialism was the negation of capitalism, and capitalism relied on the market, state provision in place of market provision became the defining characteristic of socialism in practice. This was common ground between the realists, who understood that a powerful state would be a permanent feature of socialism so conceived, and the Utopians, who still clung to the original vision of a communist society beyond the state.

But the equation of socialism and state provision was misconceived. Marx had an inkling of a better view when he saw that the point was not to negate capitalism but to transcend it, which meant taking over and preserving the valuable elements in that system while replacing those that had become historically outmoded. As we have seen, however, in his description of communism the valuable elements were narrowed down to the material achievements of capitalism: its technology, its machinery, its human skills. Marx failed to address the question as to how these achievements could be preserved without the extensive use of markets as a necessary element in an advanced industrial economy. At a deeper level, he failed to ask how it was possible in post-capitalist society to reinstate a form of community that was characteristic of simpler, less dynamic modes of production.

I have suggested so far that socialist hostility to markets goes farther than simple opposition to the market rhetoric of the New Right. Anti-markets attitudes are deeply embedded in the socialist tradition for historical reasons. I have also suggested, however, that this tradition contains an internal tension, perhaps even a contradiction. It wants to make available to the mass of people the enormous benefits that industrialization, predominantly in its capitalist form, has achieved, while at the same time remaining romantically attached to a pre-industrial vision of community. If we are to begin reworking the philosophy of socialism, we must be prepared to face squarely up to that tension and discard components of the tradition which closer analysis reveals to be unattainable. In the following section I give some reasons why socialists (and more generally those on the Left who espouse the values on which the analysis rests) should look favourably on markets. I then explore some elements in the traditional socialist critique of markets and ask how destructive they are of the market socialist proposals advanced in this book.

THE CASE FOR MARKETS

The values I shall appeal to in defence of markets are welfare, freedom, and democracy. An important aspect of welfare, though certainly not the only aspect, is material well-being, as measured by the quantity, quality, and range of goods and services available to people. As we have seen, an important part of the nineteenth-century socialist case was simply that capitalism left most people badly off, despite having the productive potential to enrich their lives. The question then is whether the way to increase material well-being is to replace markets generally by other forms of provision, or instead to redistribute resources so that markets bring about a more equal distribution of welfare.

The case for markets here is familiar, but it bears summarizing. Markets serve simultaneously as information systems and as incentive systems. The price mechanism signals to the suppliers of goods what the relative demand is for different product lines, while at the same time giving them an incentive, in the form of potentially increased profits, to switch into lines where demand is currently high in relation to supply. These two functions are separable, a point which is worth underlining. Even if we imagine people so socially responsible that they require no private incentives to employ themselves in the most useful way, there is still need for a mechanism to signal what that most useful employment is. Profits do just that, although the good citizens we are imagining would be happy to hand them all over to the community chest, once their function had been served. The point to stress is that, for markets to operate effectively, individuals and enterprises must receive primary profits, but the proportion of those profits that they need to keep as private income depends on how far they require material (as opposed to moral) incentives. The market is flexible in this respect, accommodating perfect egoists and perfect altruists as well as the majority of us in between who need some material reward to motivate us, but are happy to contribute a proportion of our takings to the public purse.

If we forgo the signalling function of the market, we must look for some other way of co-ordinating our behaviour as producers of goods and services with our demands as consumers. The early socialists seem to have thought that what needed producing was so obvious that only informal co-ordination was required. Perhaps in a small community with a very simple style of life this might be so. But if we think of a large industrial society producing an enormous range of goods and services, the only feasible alternative is state planning. Now planning can take a number of different forms, and in some of these it will serve to complement markets rather than to replace them. (This issue is dealt with extensively in Chapter 5 below.) But let us consider what would be involved in the outright substitution of planned for market provision. A planning agency must allocate labour to enterprises, tell each enterprise what to produce and in what quantities, and price the resulting goods and services so that supply matches demand. There has been a lengthy debate (the so-called 'calculation debate') about whether the problems this poses are theoretically soluble. Even if they are, practical experience reveals that planned economies in fact have great difficulties solving them. The more advanced the economy, the greater the magnitude of the task, because, as the range of products increases (and the products themselves become more sophisticated), it becomes progressively more difficult for planners to control production in a way that will meet consumers' demands. Hence the familiar experience from Eastern European economies (usefully summarized in Nove, 1983) of the over- and under-supply of many items, of poor quality goods, and so forth.

It would be a large over-simplification to suppose that personal welfare can be measured simply in terms of the quantities of private-consumption goods that people enjoy. It depends also on such things as access to public facilities (bus services, swimming pools, theatres) and, more intangibly but no less significantly, on general features of their society such as the quality of the physical environment. These are goods that markets are generally not competent to supply. How the various components of welfare are to be added up is a matter for each individual to decide. Some socialists seem to have a tacit moral preference for collective consumption over individual consumption, but in general this is indefensible. There is nothing intrinsically more desirable about swimming in a public pool than swimming in a private pool; the case for public pools has to be made out in terms of comparative efficiency. Since tastes differ, there will be many goods and services that are better supplied as private-consumption items, and here the case for markets comes into its own.

From this perspective, the major criticism of capitalism is that it distributes welfare too unequally, by failing consistently to find employment for everyone who wants it, and by generating excessive income differentials between different groups of workers and between owners and employees. Market socialists aim to rectify these defects, first by the public regulation of investment to ensure full employment, second by encouraging the growth of forms of enterprise (especially workers' co-operatives) in which primary income is distributed more equally, and third by using the tax system to implement such further measures of redistribution as command general assent. Again, the details are not germane here. The general point is that a market is a highly efficient mechanism for controlling the production of goods and services, but that the distribution of welfare that a market generates depends on the framework of public institutions—property rights, investment agencies, tax systems—which surrounds it. Rather than dispensing with markets, market socialists want to change these institutions radically so that the material components of welfare are spread more evenly throughout the community.

Freedom as a value has recently returned to prominence on the Left, as socialists have begun to realize how politically disastrous it is to allow the New Right to equate the free society with capitalism. A socialist view of freedom centres on the idea of effective choice: a person who is free has many options to choose between, but these options must be real rather than merely formal. This implies that freedom can be diminished not merely by legal prohibitions but also by economic policies that deprive people of the material means to act on their choices. The degree of freedom in a society is closely connected to the way in which it distributes its resources. But it would be a mistake to conclude that equal distribution by itself, in the absence of an effective system of exchange, is enough to maximize freedom. For what if the resources that are equally shared are resources that nobody wants, or if there is uniform provision that suits some people but not others? Freedom is valuable precisely because of the possibility that people may make radically different choices about how they want to live their lives.

Markets, therefore, must have a central role to play in a society that aims for freedom, for they allow people to choose the resources that suit their particular styles of life. People can dress as they please, pursue their particular tastes in music, and so on, provided only that some supplier responds to market incentives and delivers the necessary goods. It is easy to slight these freedoms: no one is going to defend the intrinsic value of punk hair-styles or designer jeans, but this misses the point. The value resides not in the particular choices that people may make but in the capacity to choose, and the sense this gives them that they can define their own social identities—outrageously if they like. Nobody wants to have to justify choices of this kind to some public agency, however democratically constituted.

Besides this general freedom of choice which markets confer, there are specific freedoms whose value is worth emphasizing. One is the freedom to choose a type and place of work. In no society can this freedom be unlimited, since ultimately the work that people do must be matched to the goods and services that need to be provided, but in non-market systems labour must be directed by command, persuasion, or something similar. Market economies rely on financial incentives, and this has a number of advantages. It caters for the person who wants to work in an idiosyncratic way for a low income—the proverbial artist starving in a garret—and also for the highly skilled person who for one reason or another decides not to make the socially optimal use of his skills. In both cases, the choice is placed firmly in the hands of the person in question: either do X and earn a larger income or do Y and enjoy benefits of other kinds. In a system of labour direction, such people will be at the mercy of the directing agency who may or may not listen to appeals to be allowed to do Y. The idea that choice of work could be fully voluntary in a non-market system—not even governed by exhortation and social pressure—is too fantastic to be worth contemplating.

Another freedom that requires the existence of markets is political freedom, in the sense of the freedom to express and communicate political opinions, possibly widely at variance with the view of the majority. Obviously markets cannot of themselves guarantee that this freedom will be preserved; political censorship may prevent that. But on the other hand it is difficult to see how freedom of expression can be realized unless there is a market in books, newspapers, and such other media of communication as technology makes available. People must be permitted to band together to publish their opinions; if they are not allowed to sell the results, they must be funded by a public agency—and how is the agency to judge which publications are worth supporting? Almost inevitably the dominant political view will affect the way this question is answered.

One does not have to be a fan of Mr Murdoch and Mr Maxwell to see the force of this argument. In a capitalist economy communications tend to be dominated by a few giant corporations, whose bosses have excessive power to mould public opinion. Freedom would be enhanced if there were more competition, if readers and viewers were exposed to a wider variety of sources. A socialist market economy ought to move in this direction, encouraging the growth of many small enterprises publishing books, periodicals, and newspapers, making television programmes and films, etc. The structure of the market depends on investment policy, and under market socialism this can be formulated so as to encourage freedom of expression.

What if the policy is less enlightened? A fear sometimes expressed is that, since investment is a public function, dissident groups may find themselves starved of capital by the agencies that supply it. Our precise answer here will depend on how the investment agencies are constituted. Two general points are, however, worth noting. First, whatever their precise constitution, the agencies' task is to allocate capital on the basis of the commercial viability of enterprises, modified by such social considerations (regional employment needs, etc.) as are written into their mandate. To discharge this brief, there is no need to make judgements about the intrinsic value of what is produced, any more than there is when deciding, say, whether to invest in a co-operative making plastic flowers. Second, even if political pressure is exerted to deny the dissident collective funds, it is likely still to be possible to publish in some form, given a market context. Equipment can be leased, paper bought, and the product sold—no doubt with ringing denunciations of the political bias of the investment agencies on the first page. Complete suppression is very unlikely without overt censorship, which is possible in any type of economy and can only be combated by direct political means.

A socialist market economy, then, can reasonably promise freedom in choice of personal consumption, freedom in choice of work, and freedom of expression. It aims to extend these freedoms more widely than is possible in a capitalist economy; and a socialist economy without markets, or with very limited markets, would be unable to guarantee them. Freedom is always threatened by discretionary power—that is, the power of one agent to determine whether a second shall enjoy some benefit—and this is true whether the power in question is the power of the traditional capitalist over the working lives of his employees or the power of the socialist official over the recipients of state benefits. We cannot do without discretionary power altogether (which is to say that freedom must sometimes make way for other values), but we can limit its scope by extending people's entitlements to resources which they can exchange for other resources according to choice. Markets are essential to that process.

The idea that markets contribute to freedom is familiar, not least from the distorted form which this idea takes in the writings of the New Right. The idea that markets may contribute to democracy is perhaps more novel. Again, there is a libertarian version of this thesis that I want to reject. This holds that the market is a kind of permanent plebiscite in which the consumer registers a vote each time he or she purchases one item rather than another. The flaw in this argument, easy enough to spot, is that the number of 'votes' a person has to cast depends directly on his or her income, whereas democracy is supposed to be a system of political equality. Even though market socialism aims to allocate income far more equally than capitalism, it would be egregious to represent it as democratic on these grounds alone.

Instead my focus will be on two more familiar species of democracy: industrial democracy and democracy in the state. For industrial democracy to be meaningful, the members of each enterprise must have a substantial degree of control over their work environment, including decisions about the range of products to be made, the method of production, and so forth. In a market economy they are able to have this autonomy. They are of course constrained by prevailing market conditions; they may find it financially impossible to supply the good or service that they would ideally prefer. But, except in cases where the workers have very specific skills or the available machinery is unadaptable, there will be a range of options to choose between. Enterprise members can debate whether to specialize or to diversify, whether to go all out for maximum production or to opt for more pleasant working conditions and a lower income. Decisions such as these are the stuff of industrial democracy. If, on the other hand, the economy is fully planned, no enterprise can enjoy comparable autonomy. Each must be given input and output targets which between them largely determine the remaining aspects of the enterprise's work. There may still be minor matters over which democratic decision-making is possible—the shape of the working day, for instance—but the experience of workers will primarily be that of executing orders handed down from a central planning agency.

Markets permit industrial democracy but do not necessitate it. The extent to which it is possible depends on the structure of enterprises, with workers' co-operatives being the most democratic form. Such an enterprise structure may not be optimal for all industries at all times, in which case we face a trade-off between economic efficiency and industrial democracy. Even so, a socialist market economy would provide a democratic work environment for those who valued this most highly, on the (reasonable) assumption that not everyone would share such a strong preference.

A supporter of socialist planning might concede that individual enterprises could not be democratically self-governing in a planned economy, but argue that this would be compensated for by the fact that the economy as a whole would be subject to democratic control; a market, in contrast, leads to a competitive outcome that no one controls—the survival of the fittest. I should like to suggest that the opposite is true.

To begin with, we must recognize that any modern state that aspires to be democratic faces a major problem, namely how to ensure effective control of the specialist administrators who man the various branches of the bureaucracy. In its simplest form, the problem is that these experts have access to bodies of information which—quite apart from the issue of government secrecy—no ordinary citizen could possibly be expected to absorb. The proposals coming from the bureaucracy are in these circumstances hard to challenge with any show of conviction. Now, in so far as this problem has a solution, it must reside in the citizen body laying down general guide-lines for the bureaucracy to follow, rather than in trying to settle matters of detail. To give a simple illustration, the public might decide whether it wants nuclear power stations or not; if the decision was positive, the experts would then decide which type of station to build in which area. It is immaterial here whether we see this process as occurring through the representative institutions that we now have, or through something more radically democratic (primary assemblies or referendums, for instance). The point is that realistic democratic control depends on being able to make such a separation between general questions of principle and specialist decisions.

Economic planning, of the type attempted in the Soviet Union and elsewhere, tends to obliterate any such distinction. A viable plan must be both comprehensive and detailed; it relies on gathering a vast amount of information about the current state of the economy and projecting into a future time period. This requires a large bureaucratic machine whose task is to co-ordinate the information to the best of its ability. It would be virtually impossible for an outsider to challenge the resulting plan, since, if one element is changed, this has repercussions for all other variables. The essence of such a plan is its connectedness. 'Democratic planning' is for that reason a mere slogan. No democratic body, composed of non-specialists, could possibly assimilate enough information to draft a plan of this full-blooded sort.

Framework planning, the kind of planning which market socialism requires, is a very different matter. Here, instead of trying to determine the detailed shape of the economy, we are simply laying down broad parameters within which the economy will find its own equilibrium. At stake here are issues such as preferred enterprise structure (if any), guide-lines for

investment agencies, optimal tax rates, and the like. These
issues are by no means easy to settle, but at least they are issues
of the right kind of generality for debate in a democratic forum.
Precisely because the role of the state is restricted under market
socialism, it becomes possible to contemplate effective democratic control of the bureaucracy. No democratic body can
reasonably decide what next year's quota of Cheshire cheese
should be, but in a socialist market economy this is not a
question that anyone need decide; the makers of cheese will
adjust their supply week by week to match the demand, and at
the year's end whoever is interested can find the answer.

In this part of the chapter I have offered arguments in support
of markets under socialism—arguments appealing to the values
of welfare, freedom, and democracy. Along the way I have
suggested reasons why market socialism might score more
highly in these terms than contemporary capitalism, but my
main purpose has been to rebut the view that socialist ends are
best served by statist means. I now turn to some critiques of
markets which draw on traditional socialist ideas, and again I
shall isolate three elements for separate discussion.

SOME CRITICISMS CONSIDERED

The critiques in question are often expressed programmatically
in terms of the importance of meeting needs rather than wants,
or in terms of the contrast between production for use and
production for profit. These slogans roll together the three
elements I want to separate. The first claim is that a market
economy produces the wrong goods and services: it responds to
superficial demands as opposed to real needs. The second is that
the goods and services so produced are distributed in a morally
arbitrary way. The third is that a market breeds selfish motives
in both producers and consumers, so that the general quality of
relationships in the society is corrupted. Although there are
obvious connections between these three claims, they are by no
means identical, and indeed it is essential to keep them separate
if we want to evaluate them properly.

The first critique can be seen as a way of undermining the welfare argument for markets given above. A market may be a highly efficient way of producing a large range of goods and services, but that is all to no avail if the goods and services do not conduce to genuine well-being. For the critique to carry any weight, however, we must have some way of identifying 'real needs' in contradistinction to the demands actually revealed in consumer behaviour. How is this to be done?

One possibility is that the critic is simply postulating a universal list of human needs, grounded in a theory of human nature. Now the idea of universal human needs is not entirely bogus: clearly there are prerequisites which all of us must have to survive and flourish, such as adequate food, protection from the elements, and so forth. But such a list is not going to be very extensive, and, more important, the more affluent a society becomes, the smaller a proportion of its output will be required to cover these items. Judged in these rigorous terms, most of the things that people consume in advanced industrial societies are non-essentials. If the critic tries to extend the list of universal needs beyond the basic essentials, he can be charged with arbitrarily attempting to impose his preferences on others who do not share them. For the fact is that people have very different ideas about how life should be lived beyond bare survival, and so their priorities in terms of the goods and services they want to have also vary greatly. Rather than attempting to impose spuriously uniform needs on individuals and societies, we should be trying to create an environment in which the most diverse styles of life can coexist harmoniously.

There is however a second, and less obviously flawed, way of presenting the critical argument. The claim is now that the desires revealed in market behaviour are to a large extent induced by the producers of goods and services themselves, who have an obvious interest in stimulating demand for their products. 'Real needs' are the desires people would have if they were allowed freely to make up their own minds without such stimuli. In this version, there need be no presumption that everyone's real needs are the same.

There is obviously much truth in the view that market demand is externally stimulated. But it is much less obvious why this should destroy the welfare argument for markets. For personal well-being is not simply a matter of having as few unsatisfied desires as possible. If this were true, the happiest man of all time would be Diogenes in his tub. Well-being can be increased by cultivating new desires, and inevitably many of the desires that we do acquire come to us from our surroundings. Our tastes in food change because of recipes other people try out on us, and we learn our clothes sense by looking at what our neighbours are wearing. There is nothing sinister in this, and anyone who finds fault with it must condemn us all to solitary confinement. Why, then, should the case be any different with desires that are stimulated by those who, as producers, have a vested interested in doing so?

The difference can only be that these desires are stimulated deliberately, and, because of this, may sometimes be aroused in ways that undermine the consumer's autonomy. It is the means of stimulating, not the fact of stimulation, that is crucial. In the great majority of cases in which new desires are created, there is no threat to autonomy. A desire may, for instance, merely be instrumental to the satisfaction of pre-existing desires, as when we choose a product which we believe will serve an existing purpose more effectively—say a spin-drier in place of a hand-wringer. Or we may simply acquire a new taste, as when we try a new fruit that has found its way on to the supermarket shelves. The changes of desire here may not be deliberate on our part, but there is nothing in the process of acquiring the new wants which, if it were brought to light, would cause us to renounce them. Moreover, although the suppliers of spin-driers and pomegranates have no direct interest in our welfare, but only in selling their products, it does not follow from this that our welfare cannot be increased by responding to their promptings.

The cases which do cause concern are those which involve some sort of failure of rationality on the part of the consumer. The most clear-cut examples—which are also the least worrisome—are those in which the producer dupes the consumer into believing that his product will do things which it will not, in the most literal sense. Obviously we must have consumer information services, trades description acts, and the like to counter this possibility, and these will need to be built into the political framework that surrounds a properly functioning market. But note too that consumers will sooner or later become aware of their errors—when the can-opener fails to open their tins—and will usually be in a position to choose better in the future. (Where this is not so, for instance where a bad choice may cause permanent damage—as in the case of medical care—there is a strong case for providing the good or service in question outside the market.) So markets function as learning devices in cases where products are bought repeatedly, and failures of rationality of this kind tend to be self-correcting.

More disturbing are those cases in which the claims made by the supplier of a product are intangible, but none the less persuasive: the product is portrayed as essential to the good life, or social success, or success with the opposite sex. Claims of this kind are not only largely unverifiable, but inexhaustible: perhaps you now drink the jetsetter's aperitif, but are you wearing the right shirt while doing so? Critics of the market economy batten on to these cases, pointing out that markets may produce a profusion of ever more costly goods whose real contribution to the increase of human happiness is nil. Everyone is caught up in a scramble for commodities where, as soon as one level of consumption is reached, a new ascent is called for in search of the elusive Shangri-La.

We need to be reminded from time to time of this uncomfortable truth. But where does it lead us? I am not convinced that it weighs decisively in the choice between market and non-market methods of provision. If anything, it cautions us against exaggerating the contribution of economic arrangements of any sort to human welfare at the deepest level. Sages from time immemorial have told us that real happiness depends on self-knowledge, on good personal relationships, on adjusting your aspirations to your capacities, and so forth. Material standards do not make so much difference—at least once basic needs are satisfied. If we accept this wisdom (which most of us do in theory, though rather few in practice), the observation that markets increase material levels of consumption far more than real happiness will seem less telling. For the same is likely to be true of all economies operating above the sub-sistence levels. Once non-essentials are put into production, people will begin demanding them for psychologically suspect reasons. We can see this only too clearly in the case of Soviet-style economies, where the scramble for commodities is every bit as intense as its counterpart in the West. It does not take self-interested producers with sophisticated advertising techniques to create this psychology. Rather, because the psychology exists—the illusion, if you like, that happiness can be bought through commodities—producers can make use of it to sell particular items.

We judge any economy too harshly if we ask how much it contributes to inner happiness. The case for markets is that they are an effective means of supplying many goods and services to consumers. Some consumers will find that what they buy fails to live up to their expectations, but the worst that can be said about markets is that they reinforce, rather than challenge, the psychology that brings about this result. Short of a wholesale onslaught on the private consumption of goods and services—a remedy so drastic that even the most ardent socialist will surely shrink from it—the problem is insoluble. Education can help to mitigate it, by getting people to think more clearly about what they want out of life, but the psychological processes involved in consumption of this kind are too complex ever to be brought fully under rational control.

Having peered briefly into this abyss, let us turn our attention to the alleged distributive failings of the market. A well-established line of attack is that markets distribute goods and services not according to need but according to ability to pay, which in turn depends on success in a kind of social lottery. The first part of this claim is obviously true, in the sense that, although people can use their available purchasing power to meet their needs for particular items, there is no general reason to expect the extent of purchasing power to match the extent of need; often the reverse is true, as in the case of handicapped persons who have greater-than-average needs but less-than-average capacity to earn income. Markets must be corrected by distributive mechanisms that take account of this fact—though whether this should take the form of a redistribution of primary income or the provision of goods and services on a non-market basis is an issue to be decided on a case-by-case basis.

A more difficult question is to decide how extensively 'need' should be construed. I suggested earlier that the list of universal, basic human needs is really quite short. Over and above that, there are needs that are defined as such in the context of a particular society: there is a general consensus on the level of provision that each person ought to enjoy if they are not be excluded from normal social life (for evidence, see Mack and Lansley, 1985). Needs of this latter sort should certainly be met, but there is no reason to think that they will come anywhere near exhausting a society's resources. So distribution according to need cannot be a complete distributive principle, in the sense of a principle that can sensibly be used to allocate all the available consumption goods.

Take housing as an example. Housing is a need in the sense that everyone needs accommodation that comes up to certain (socially determined) standards—so much floor space per person, running water and drainage, adequate heating. But 'need' in this sense could not be used to allocate the whole of the housing stock, since it sets only minimum standards, whereas most accommodation offers additional benefits that will be valued differently by different people. Some people prefer old houses of character, others prefer modern dwellings that are easy to run, etc. There is quite properly a housing market, circumscribed by the obligation of political authorities to ensure that everyone has access to housing that meets the minimum standards. This is often the way that needs intersect with market provision. Markets are allowed to operate, but subject to political supervision aimed at ensuring that all needs are met.

What of the second part of the critical claim, that market allocations of income are morally arbitrary? Income distribution depends on the background institutions against which the market operates: rules of property, rules of contract, tax rules. As I have already argued, the extent of inequality in a socialist market economy will depend on the shape these institutions take, which in turn depends on the strength of political will in favour of redistribution. We should assume, however, that to some extent income will depend on success in market competition—in most cases success that is shared among all the members of a particular enterprise. How morally arbitrary is this?

It is not morally arbitrary if success can be shown to be deserved. Desert can be looked at in two ways. First, it may be narrowly tied to the idea of voluntary choice and activity: what you deserve depends on what you have chosen to do. In a market context, decisions about how long to work, how hard to work, what product to make, what method to use, what new skills to acquire, can all be seen as relevant to desert in this narrow sense. In so far as an enterprise's success can be attributed to factors such as these, its members deserve their additional income. If, for instance, they react to excess demand for a particular product by extending their hours of work or by reorganizing jobs to increase output, their social contribution has increased and they have earned their extra rewards.

Second, desert may be extended to cover natural talents and abilities and other skills not voluntarily acquired. This extension is a matter of philosophical dispute: some would argue that we can only deserve on the basis of our voluntary choices and actions; others that we can deserve on the basis of what we are, even if our personal characteristics are to an extent non-voluntary. The practical difficulty this poses is that markets do not discriminate between the two cases: they reward the naturally able as well as those who have chosen to cultivate certain talents. Nor does there seem to be any method of taxing away the returns of natural ability that is both feasible and fair. We are therefore likely to feel most comfortable with the results of market distribution if we take desert in its wider sense. To those who are wedded to the narrower interpretation, I offer the following question: is there any alternative system of distribution that will come closer in practice to the ideally just distribution, namely one under which receipts depend entirely on voluntary choices?

We still face the objection that market incomes do not depend solely on desert even in the wider sense, but to a degree on luck. An enterprise may make a large profit if a new product which it launches turns out to be a best-seller, although there was no real reason to believe beforehand that this would occur (obviously we must separate genuine luck from careful market research, which falls under the desert principle). The role of luck cannot be denied, but the key question, in my view, is whether the background institutions against which the market operates tend to consolidate luck or to disperse it. Market socialist institutions, under which windfall gains are shared throughout enterprises, and successful enterprises are not able to multiply their gain by investing the proceeds in other enterprises, will tend to disperse it. Since what matters from the point of view of fairness is not income in any particular time period, but lifetime income, the system might have some of the features of a genuine lottery in which punters win on some rounds and lose on others, the net effect being relatively insignificant. I take it that the socialist objection is not to luck of this sort, but to the kind of luck which, once enjoyed, puts its beneficiary into a position of permanent advantage.

To sum up, markets should be seen as working alongside other institutions whose aim is to redistribute in line with need. Some part of the inequalities they generate can be justified on grounds of desert; and we can attempt to neutralize the remainder by making the market into a genuine lottery, not a game of cumulative advantage. Even if markets do not match our distributive ideals perfectly, a properly framed market may approximate as closely to those ideals as any other system will in practice (we need to think realistically, for example, about the problems of distributive justice in bureaucracies).

We come to the third and final part of the traditional socialist critique of market economies. This is the aspect which most acutely exposes the tension noted at the start of this chapter, between modernizing and backward-looking elements in socialism. Markets are faulted for their competitive character, for dividing people instead of uniting them in community. The emphasis here is not on the economic performance of markets, but on the quality of human relationships that they foster.

Two preliminary points of clarification should aid discussion of this claim. First, it should be abundantly clear from what has been said here and elsewhere in this book that advocates of market socialism do not regard markets as the sole mechanism by which people should be related in a socialist society. Markets are seen as indispensable for economic purposes, but they should be complemented by democratic political institutions, by planning agencies which set the parameters of the market, by publicly funded social services, and by a voluntary sector in which altruistic concern for others can be expressed directly (e.g. in community programmes of various kinds). In other words, people would be linked together in a variety of ways in the society that is envisaged. Second, I have pointed out already that the economic case for markets does not depend on any underlying assumptions about human nature; in particular we do not need to assume that people are inherently selfish. All that is required is that people in general should display economic rationality—that is, behave economically in such a way that the net value of their holdings is maximized. It does not matter if their underlying motive is simple greed, or a wish to confirm Divine election (as Weber supposed), or a desire to benefit their fellow men.

Once these points are borne in mind, it becomes apparent that markets are incompatible with communitarian relationships only if community is defined in a very strong way. It must be being regarded as a form of association that is all-encompassing, both in the sense that each member's relationships are all and only with fellow-members and in the sense that these relationships are all of the same character. I shall call such a community monolithic. Monolithic communities exclude markets because they leave no space for the instrumental behaviour that markets require—if I must always make the welfare of others my direct intention in acting, then I cannot barter and exchange with them even if I believe that behaviour of this kind would indirectly maximize the community's welfare.

But merely to spell out this strong view of community reveals both its unattractiveness and its implausibility. Monolithic communities are closed societies which deny individuals space to develop their own personalities and styles of life. They were characteristic of pre-industrial epochs—though it is questionable whether even the celebrated village communities of those times were quite as monolithic as we now tend to think. In any case, we should not wish to revert to them. Our preference must be for a looser form of community which allows us space to develop as individuals—as well as to contract, if we wish, into more intense communities (monasteries, communes, etc.). But this immediately makes room for markets as devices which can link together people and groups whose relationships are communitarian only in this looser sense.

The strong definition of community is also implausible. It supposes that the quality of relationships which makes community valuable is all-or-nothing. But in fact even our most intense relationships seem able to withstand a much wider range of role-playing. Consider friendship. People can remain close friends even though they find themselves at times competing with one another—say in the market-place or on the sports-field—or on opposite sides in a political dispute. Multiple role-relations of this kind pose practical dilemmas—how hard can I compete with Jim without jeopardizing our friendship?—but in real life people are constantly meeting and resolving such dilemmas successfully. Why should it be any different with community? I may identify strongly with a group of people and take a deep interest in their welfare, while on some occasions finding myself in conflict or competition with particular members. We are sophisticated creatures who do not find it unduly paradoxical that we should play different roles in relation to one another in different aspects of our lives.

So far I have been arguing that markets and community may be compatible with one another. I have not tried to show that the market economy is itself a form of community, a view that I find implausible precisely because community does depend on the intentions that we have when we interact with others. We need institutions alongside the market to embody our communitarian commitments: the institutions of politics itself, the public services, the voluntary sector. These are all avenues in which we can express, individually or collectively, our concern for the welfare of other members of society.

But might not markets inhibit the growth of such concern? Whatever their formal requirements, do not markets *in fact* tend to encourage people to see themselves as self-sufficient individuals whose motto is 'Chacun pour soi et Dieu pour tous'? These residual anxieties are often expressed by critics of market socialism. They cannot be finally resolved until we have a working model of such an economy in which to test the implicit psychological claim. What can be said, speculatively, is that the cult of individual success seems to be a specific by-product of capitalist markets where the success of enterprises is easily identified with the achievements of the individuals who run them. With collectively owned enterprises—say workers' co-operatives—one would expect instead that the qualities found most valuable, and extolled in the popular mind, would be teamwork and contribution to collective endeavour.

CONCLUSION

In the opening section I drew attention to the historic tension in socialist thought between a modernizing commitment to industrial society and a nostalgic attachment to pre-industrial forms of community. In resolving that conflict, we have had to discard such outmoded visions of community and replace them with a different understanding of what community can mean in a modern industrial society. This revised view makes room for market relationships, although it recognizes that people must be linked together in other ways as well if they are to realize themselves fully as social creatures.

The tension must, in my view, be resolved in this direction. A politically viable form of socialism must base itself on the aspirations that people actually have, and this means people whose experience has been shaped by a century and a half of industrialization in its capitalist form. A great deal of weight is attached to personal independence, to having a style of life that suits your own particular tastes and inclinations. Too much weight, some socialists might reply: there ought to be a greater sense of interdependence and social responsibility. Even if we share that belief, we cannot solve the problem by compulsion, by corralling people into monolithic communities. We must find social institutions that respect their independence, while at the same time providing channels through which social concern can be actively expressed.

Markets must, therefore, play a large part in a feasible form of socialism. The reasons are not merely those of economic efficiency (important as these are), but also those of diversity and personal freedom. The evidence we have suggests that people like having cash in their hands and buying their goods and services competitively: they feel secure and self-confident in a way that they often do not when dealing with public agencies. This is the promise and attraction of capitalism, but in all too many cases it is nullified by a gross maldistribution of resources. So we need institutions outside the market itself—political institutions, primarily—that will set a new framework within which the maldistribution can be rectified. Market socialism involves neither a simple-minded endorsement of markets, nor their straightforward dismissal, but instead a discriminating response that tries to do justice to the complexities of human nature as we see it displayed around us.

3

Socialism, Markets, and End States

Raymond Plant

IN this chapter I want to consider the degree of compatibility between markets and traditional socialist values. Clearly there is no point in market socialists putting forward market-based views if such views cannot be reconciled with accepted socialist values. A failure to show at least a reasonable degree of compatibility would lay us open to the charge that market socialism is a contradiction in terms. It is very important that socialists who are attracted by the market as a centrally important institution of a free and productive society should be in a position to distinguish between good and bad arguments in favour of markets and indeed arrive at a proper characterization of their properties. In recent years the New Right has taken the initiative in political debate, particularly in relation to the role of markets, and it is vital that in coming to endorse a role for markets socialists do not accept uncritically the account of their role adopted by neo-liberal theorists. This is particularly important in the context of understanding the relationship between markets and typical socialist values such as social justice. I shall discuss primarily the values of freedom, social justice, needs, and community in relation to markets and also say something about the traditional role envisaged for planning in realizing such values.

PROCEDURES AND END STATES

One of the central themes which I shall try to develop is the relationship between procedural and end-state principles in political thinking and the extent to which markets are usually envisaged as embodying the former and socialist values the latter. The distinction can perhaps be put in the following way. It might be argued that traditionally socialists have been interested in particular social outcomes, for example greater equality, the satisfaction of a wider range of basic needs, the achievement of greater effective liberty for all citizens, the development of citizenship as a positive status with guaranteed rights to positive resources such as income, health care, and education, and a greater sense of fraternity and community. Socialism is usually defined in terms of such ends in varying clusters and combinations so that the socialist aspiration is towards a particular end state of society in which these values will be achieved. It is therefore a goal-directed theory. Other aspects of socialist belief are frequently seen as means towards these ends: nationalization and public or social ownership, for example. Indeed some of the disputes between revisionists and Marxists have been over the extent to which certain means such as nationalization are necessary to the realization of socialist ends on which, it might be argued, both protagonists are agreed— Marxists arguing that greater social justice and a more communitarian social order cannot be achieved without the common ownership of the means of production, and revisionists arguing that common ownership, because it has the status of a means, has to be a contingent feature of socialism, with any proposal for common ownership of particular industries being assessed in terms of its likely effectiveness in achieving socialist goals. On this view common ownership is not the essence of socialism but rather a potential means, the merits of which have to be considered in particular circumstances in relation to the realization of socialist values. In this sense, therefore, socialism is an end-state doctrine to be defined in terms of an aspiration to the achievement of particular goals which, in the socialist view, can be made determinate in both theory and practice.

In contrast to these goal-directed and end-state theories characteristic of socialism, markets seem to be paradigmatically procedural institutions in which no particular outcome in terms of the distribution of resources can be expected, at least in so far as the market is allowed to operate freely, independent of government regulation. This is certainly the position of current neo-liberal defenders of the market such as F. A. Hayek. This arises for the reason, going back to Adam Smith, that markets, while obviously the result of human action, do not produce results which are the products of human design. So, for example, in relation to distributive justice the neo-liberal will argue that the categories of social justice are irrelevant to the market because the market is a procedural and not an end-state institution, just because its results are unintended, undesigned, and unforeseen. The market distribution is an unintended consequence of individual actions and exchanges which were undertaken for all sorts of different reasons. Certainly some people end up with more and others with less, but this is not a distribution in the sense that the socialist typically wants to talk about the distribution of income and wealth, with the assumption that it is at least potentially maldistributed and in need of correction. It is rather an unintended distribution and the market as such is a procedural institution which is indifferent to any substantive end state whether in terms of social justice, equality, effective freedom, or community. It is, to use the apt description of Fred Hirsch, 'in principle unprincipled'. Given this understanding of the market mechanism, it is not surprising that many socialists have seen a deep incompatibility between the market and socialist values: the former procedural and indifferent to outcomes; the latter substantive and defining its vision in terms of particular end states.

However, the issue goes deeper than just the contrast between procedural and substantive values, because on the face of it the two could be run together in the way to which we have become accustomed during the period of Keynesian social democratic consensus after the war in most countries of Western Europe. This was the view that the market should be allowed to operate within a framework, determined by the government, within which certain sorts of substantive outcomes were to be secured, either through government intervention in the market or through government providing goods and services as the result of market failures. In this way, the government could seek to make the market responsive to social goals such as greater social justice, equality, and full employment. The government could also provide in a predictable and universal way for all, goods which the market would be unlikely to produce, such as rights to welfare goods of various sorts. However, this combination of the free market plus welfare spending in the pursuit of socialist goals such as greater equality and justice has not only become very difficult to maintain in practice, but also creates, in the view of the liberal market theorists, deep theoretical difficulties in attempting to graft a particular patterned outcome on to a procedural mechanism such as the market. This is because in the market patterned or end-state principles are in fact defective and these theoretical defects make such substantive goals impossible to achieve in practice: a fatal conceit, to use Hayek's phrase. The defects in the substantial principles of socialism are, in the liberal market theorist's view, intimately connected with the character-ization of markets which I have just outlined. Those values such as freedom, justice, and equality which the socialist espouses and which, on that view, seem to require particular outcomes in terms of the ownership of property and entitlement to income and welfare are given a negative and not a positive interpretation by the market theorist. In the liberal market theorist's view only a negative interpretation of these values will make them compatible with markets. This argument is absolutely crucial, because if correct, it would mean that market socialism would either be incoherent, running together end-state and procedural principles, or would have to produce an interpretation of socialist values which would portray them in the same kind of negative and procedural way. This would make the market socialist position indistinguishable from neo-liberalism. In my view, this issue is at the theoretical heart of the debate about the extent to which socialism is compatible with markets and the attempt to deal with it will dominate the remainder of this chapter.

VALUES AND PROCEDURES

What does it mean then to say that the market theorist treats the positive social values which are at the heart of socialism in a negative and procedural way? At the centre of this debate is the connection between the reinterpretation of social values and the characterization of markets as unintentional and unprincipled in terms of their outcomes. We shall consider the argument here in terms of two values: social justice and liberty. According to the usual socialist view, the free market is defective because the distribution of goods and services, income and wealth that occurs through the operation of the market does not secure social justice according to the usually favoured socialist criterion of the equal satisfaction of needs. Given the inevitably random element in market outcomes, those whose needs are not met by the market have a defensible moral claim on the resources of those who are successful in the market. Hence, left to its own devices, the market causes injustice, an injustice which can only be rectified either by state intervention in the market so that it does approximate in its directed outcomes to meeting needs and the demands of social justice, or by the state providing an alternative to markets via welfare provision.

The neo-liberal defence of the free market decisively rejects this argument in a way which draws very heavily upon the characterization of markets which I described earlier. The argument is as follows. Injustice can be caused only by intentional action. So, for example, we do not regard the consequences of the weather as an injustice, however Draconian its effects may be. Thousands may die in an earthquake or a flood, but this is rightly regarded as a natural disaster rather than an injustice, a misfortune rather than the infringement of justified claims. Only when the possibility of agency and intention enters does the category of injustice gain some purchase. Hence, in the neo-liberal view, agency, and in particular intention, have to be present for an injustice to have been committed. Injustice is the result of intentional action and design, not natural processes and inadvertent action.

This argument, which has a good deal of initial plausibility, is then applied to the nature of markets characterized in terms of results which are the unintended and unforeseen consequences of human agency. As we saw earlier, in markets people exchange goods and services intentionally and between individuals in such exchanges injustices may occur; for example, one individual may coerce another into an exchange. However, if we take the overall results of uncoerced exchanges in the market, we cannot regard these as unjust because they were not intended or foreseen by anyone. Hence the overall results of free markets cannot be subject to moral criticism—as they have been in the socialist and indeed social democratic and social liberal tradition—as unjust. Only individuals can visit an injustice on another individual. We must be in a position to say who has been unjust. This cannot be done in the case of the overall outcomes of markets, where the role of agency and intention becomes wholly opaque and makes markets much more like natural processes than intentional ones. Hence the proper characterization of markets as procedural institutions shows that the role of agency and intention is not sufficient to sustain a moral critique of markets in terms of social justice, as the socialist tradition would have us believe. Hence, a market socialist has either to reject this characterization of markets or to abandon his end-state view that in the interests of social justice state intervention and state supplementation of the market is justified according to this moral critique.

A precisely similar argument is used in relation to freedom. The neo-liberal market theorist accepts a wholly negative view of liberty, in which liberty is characterized as the absence of intentional coercion. I am only rendered unfree if someone intentionally coerces me. Freedom is not the positive freedom of having the ability, and hence the appropriate resources, to act effectively, but the negative freedom of not being coerced. The reasoning here involves the idea that there must be a categorical distinction to be drawn between freedom and ability, in that, if they were the same, then any kind of inability would be a restriction of liberty. There are many things which I am unable to do which it would be absurd to regard as a restriction on liberty. There are some things which I am logically unable to do: to draw a picture of adjacent mountains without a valley, for example. Other things I am physically unable to do, because of my basic physical constitution: as a man, for instance, I am unable to bear a child. I am unable to do some things because of circumstances prevailing at the time: I cannot ride up that hill today because there is a head wind. I am unable to do other things because of previous choices which I have made: as a married man with three children, I am unable to become a Carthusian monk. Other things I cannot do because I do not have the resources: I cannot go on a round-the-world cruise. It would be absurd to regard these as restrictions on my liberty. I am free to do them in the negative sense that no one is preventing me; I am just unable to do them. Freedom, therefore, should be understood, not as the possession of ability, resources, and opportunities, but rather as the absence of intentional coercion. We do not regard the wind which restricts my ability to ride my cycle as I would wish as a restriction on my liberty, because agency and intentionality are clearly lacking, as they were in the earlier examples.

These points are then applied to the case of markets understood in the neo-liberal sense. Socialists have typically wanted to criticize market outcomes in that those who are rendered poor as the result of *laissez-faire* are deprived of effective liberty, lacking the resources to act effectively. However, using the distinctions described above, liberal defenders of the market have argued that markets are not coercive in relation to the worst off, for two closely connected reasons. In the first place, as we have seen, markets lack agency and intentionality, and thus the poor who are deprived of resources are in that position not as the result of intentional action, but as the result of an impersonal process which, although the result of human action, is not the result of human design and is unforeseeable for individuals. Second, there is in any case a clear distinction to be drawn on the basis indicated above between freedom as the absence of coercion and the abilities and resources necessary for action and agency. Hence the outcomes of markets taken as a whole cannot infringe liberty. Of course, as was the case with justice, one individual in an act of exchange may coerce another, but here the agent who is acting coercively can be identified and the nature of his or her coercion identified and characterized in detail. However, this is not the case with the overall outcomes of the market. Hence again, it is argued, if we understand market processes properly, such an understanding will undercut a central feature of the socialist critique of capitalism. Again the same point which I made at the end of the discussion of social justice seems to hold here, namely, that the market socialist has either to dispute the characterization of markets and freedom offered by capitalist theorists or to abandon the idea of positive freedom or effective liberty which has contributed a good deal to the socialist justification for state intervention in the market.

There is another point that applies to the critique of both social justice and positive liberty. The neo-liberal will argue that, in so far as each is an end state or patterned principle, there is the deep and intractable problem of trying to provide a justification of the nature of the preferred pattern or end state. In the case of liberty, for example, if we define freedom in terms of the possession of abilities, resources, and opportunities, which particular abilities, resources, and opportunities are supposed to define the condition of being free? Clearly it cannot be all of the possible examples of these, because otherwise we could not be free unless we were omnipotent, that is, possessing all the powers, capacities, resources, and opportunities to do whatever we want to do. Given that this requirement is obviously absurd, how are we to decide which of these are necessary conditions for positive freedom?

The same problem applies to social justice. There are a large number of possible criteria of social justice: desert, merit, need, entitlement, etc. Clearly socialists will want to place need at the centre of moral concern, even if they find a role for some of the other criteria too, but then deep problems arise. First of all, they must provide convincing arguments to show the priority of need in relation to the others. Second, they must try to provide a determinate account of what need consists of, so as to enable a principle of need to guide policy in the distributive sphere. Third, if socialists wish to provide some role for other principles such as merit, they must decide on the relative weights of these different principles.

There are two aspects to this issue. There is, first, a philosophical problem over whether issues of this sort are capable of rational resolution. Second, there is the related problem of how, in a morally pluralist society, problems of this sort could ever be resolved in a practical way.

On the first aspect it is very important to note that many neo-liberal defenders of the market, such as the Chicago School of Friedman and the Austrian School of Menger, Mises, and Hayek, are non-cognitivists over moral questions. That is to say, they dispute the view that we can ever arrive at a rationally compelling justification of fundamental moral values. In their view, values are ineradicably subjective and attitudinal, although the Chicago School and the Austrians differ on the philosophical reasons for this value scepticism. Given this view, they dispute the socialist claim that end-state values such as social justice, equality, or need satisfaction can be given an objective moral basis. This provides them with an additional moral argument for the procedural role of the market. In a market we do not pursue some supposedly morally justified end state, but rather leave individuals who are the authors of their own values to pursue what they take to be their own good in their own way. So, for example, there can be no meaningful and compelling account of a just level of income because this would mean imposing on society one set of subjective values compared to another; rather, the only safe guide to what an individual is worth is what others with their subjective preferences are prepared to pay to obtain that person's services. In this sense markets are appropriate institutions in circumstances where we lack objective moral criteria for judging worth, merit, need, and so forth. End-state theories of socialism, on the other hand, it is argued, must presuppose some form of moral realism or objectivism if they are to be more than the arbitrary imposition of one set of subjective preferences on society. Because they avoid end states and recognize the arbitrariness of moral choice, markets are the most appropriate counterpart to ethical sub-jectivism. For this view to be coherent the socialist has either to abandon end-state values in favour of markets or to provide what the neo-liberal regards as unavailable, namely, an objective account of moral values in order to support the imposition on the free choices of the market of a set of patterned or end-state principles.

The second argument is more sociological in character—namely that in Western societies which now exhibit a wide range of moral diversity it is just not plausible to believe that there is a moral consensus available over some of the central principles of socialism, for example about needs or the degree and kinds of equality. This point has been made very trenchantly by John Gray, one of the most eloquent expositors of the neo-liberal position in Britain:

The objectivity of basic needs is equally delusive. Needs can be given no plausible cross-cultural content but instead are seen to vary across different moral traditions. . . . there is an astonishing presumption in those who write as if hard dilemmas of this sort can be subject to a morally consensual resolution. Their blindness to these difficulties can only be accounted for by their failing to take seriously the realities of cultural pluralism in our society, or (what comes to the same thing) to their taking as authoritative their own traditional values. One of the chief functions of the contemporary ideology of social justice may be, as Hayek intimates, to generate the illusion of moral agreement, where in fact there are profound divergencies in values. It remains unclear how such divergencies are to be overcome, save by the political conquest of state power and the subjugation of rival value systems. (Gray, 1983, 181)

In the view of the neo-liberals, end-state socialists are inevitably placed in this latter position. Once again neo-liberal thought in relation to markets poses a fundamental question for market socialists. If part of their own defence of markets rests upon a respect for individual preferences, why, the neo-liberal will ask, does not this respect extend to moral preferences and the consequent diversity of morals? Again, the argument is that market socialism is incoherent because it cannot consistently endorse the major moral feature of markets—namely, that individual preferences are taken as basic and incorrigible—and at the same time endorse end-state views which do not accept the outcomes resulting from the free play of preferences in markets.

An analogy will help here. The market socialist position could be likened to that of a democrat who wishes to allow the free play of individual preferences in the political sphere while at the same time reserving the right to reject the outcome of such preferences if they conflict with some end-state principles. The inconsistency here seems plausible, on the face of it, and needs to be answered in detail by the market socialist.

In the view of the neo-liberal these issues are far from being as abstract and abstruse as they might first appear. On the contrary, these problems lie at the heart of any non-arbitrary attempt to implement the socialist project. If end-state values are to mean anything, they must be capable of guiding public policy and, in the view of the liberal critic of socialism, this is precisely what they fail to do. This failure is at two levels. The first is the general level which we have just been considering, namely, that the central concepts of socialist thought such as need and justice cannot play a determinate part in guiding policy just because they are so open textured and contestable, based as they are ultimately on subjective preferences rather than objective and rational criteria. The second level of failure is more particular and immediate, in the sense that, even if we could get a consensus that, for example, medical care in society should be based upon need rather than on any other principle such as ability to pay, this principle would not help to guide decision-making in particular circumstances. Again we can turn to the work of John Gray for an example of this drawn from the sphere of medical need. Here the concept of need is likely to be as determinate and consensual as anywhere in the sphere of public provision. However, even in this case Gray argues that public officials charged with the responsibility of trying to satisfy the end-state principle of meeting need are forced to act in an arbitrary and discretionary manner just because the basic principles are far too indeterminate to guide policy.

Not all needs or merits are commensurable with each other. A medical need involving relief of pain is not easily ranked against one involving the preservation of life and, where such needs are in practical competition for scarce resources, there is no rational principle available to settle the conflict. Such conflicts are endemic because, contrary to much social democratic wishful thinking some basic needs connected with staving off senescence, for example, are not satiable. Bureaucratic authorities charged with distributing medical care according to need will inevitably act unpredictably, and arbitrarily . . . for want of any overarching standard governing choice between such incommensurable needs. . . . The situation will be the same when the occasion arises for weighing merits against each other—a process so subjective as to demand no further comment. The idea that social distribution could ever be governed by these subjective and inherently disputable notions reflects the unrealism of much contemporary thought. (Gray, 1984, 73)

Again there is a sharp rejection of end-state theories as being incapable of being implemented in a morally pluralistic society except in a dictatorial and arbitrary way. This point leads on to two further considerations: about community and about planning.

According to the views of neo-liberal critics of socialism, end-state values only make sense within a reasonably homogeneous community in which a hierarchy of ends is accepted as part and parcel of the way of life of that community. In such a society where the acids of individualism have not eaten away at the bonds of social solidarity, it may well be that there are agreed and collective views about needs and their ranking and about who deserves what in the distribution of social resources. However, we are emphatically not in that sort of position today. Even within the urban working class, which might in an advanced industrial society be regarded as the bearers of such solidaristic and communitarian values, these bonds are being broken down. Individuals and families are not bound by the solidaristic features that may have underpinned their way of life in previous generations, and that did indeed provide support for a form of socialism which in terms of its goals and values represented in a political and practical form such a way of life. However the collapse of such closed communities is not to be mourned, because they are closed and the enemies of change, mobility, and individual advancement.

Of course, this does not mean that a sense of community is valueless, but in the modern world such a sense is to be found in partial communities—groups of all sorts, within which individuals are able to pursue, in common with others ideas, interests and values which are important to them. But the point about such communities is that they are, unlike traditional, total communities, based upon choice and voluntary allegiance in a way which was not true of older forms of social solidarity. Within such partial communities there may be codes and rules of all sorts which prescribe behaviour and conventions and may embody for their own purposes end-state principles which are derived from the goals and purposes of that community— church membership is a good example. But such end-state principles are binding and authoritative only in so far as people choose to be bound by them and they are not a basis for prescribing the goals for the rest of society just because of their radical diversity. Hence, there is, in the view of the neo-liberal critic, an intimate connection between the socialist's penchant for seeing society in a goal-directed way and the value of community. According to the critic, forms of social solidarity which may at one time have supported such goals have passed away and we are left with a set of values which are not rooted in an ongoing way of life for society as a whole and which are not capable of being given some kind of rational foundation. Hence, any attempt to secure such goals and implement them is bound to be arbitrary and dictatorial. Again the free market, as we saw earlier, is the solvent of the social dilemmas caused by the failure of end-state values in the modern world. The market provides a fair procedural mechanism within which individuals will be able to pursue their own conception of the good in their own way and it will not force the realization of any particular end state on society. Community is at home in pre-industrial forms of life and it (and, the neo-liberal argues, its associated end-state principles) cannot now be grafted on to a modern complex society in which there is radical value incommensurability.

These points also apply to socialist arguments about planning. If there are to be end-state values such as meeting needs or achieving social justice, then the government clearly has to intervene in the economy in order to ensure the realization of such values. However, the market theorist will argue that any such plan is bound to be bureaucratic and radically indeterminate because, as we saw in Gray's medical example, the values which the plan would seek to implement are too incommensurable for any such plan to be realized in anything other than a bureaucratic and dictatorial way.

There are other arguments against the possibility of central planning for the achievement of socialist ends that are discussed elsewhere in this volume and they need not be rehearsed here in detail. Suffice it to say that the objections are largely epistemological. The information which planners would need to implement their plan, even assuming that it could be made determinate, is just not available in the way which would be necessary to plan in a rational manner. According to Menger, Mises, and Hayek, implicit, non-propositional forms of knowledge are necessary conditions for effective economic action in the market. This knowledge, which any economic agent has, is necessarily dispersed, tacit and relative to an individual agent. Just because it is dispersed and non-propositional, it cannot be gathered together in a way which would make it suitable for use in compiling a plan, even assuming that the goals of such a plan were sufficiently determinate, which, as we have seen, the neo-liberal denies. However, on this view planning is absolutely necessary to achieve the goals of socialism conceived in terms of the realization of a particular set of end states such as greater social justice.

All of this adds up to a formidable critique of traditional forms of socialism and demands a response. It poses dilemmas not only for traditional socialist views but also for market socialist theories: the critique, after all, is an attempt to argue a moral case for markets based upon the rejection of the socialist project. Is the market socialist committed to all or part of this sort of case for markets, and, if not, what is to be rejected? To what extent is the market socialist committed to end-state principles, and, if there is such a commitment, what effect will this have on the supposed commitment to the market?

NEO-LIBERALISM AND MARKET SOCIALISM

I believe that in its most radical form market socialism will go a long way towards accepting the neo-liberal critique of traditional socialism, based as it is upon end states and a conception of the good. The underpinning to the particular socialist element in this sort of market socialism depends upon a rejection of the neo-liberal claim that free markets are not coercive and provide a fair procedure within which individual preferences can be realized. The market socialist will argue two related theses here. The first is that, as David Miller points out, there is a radical difference between a free market as a fair procedure for recording preferences and a democratic voting system—an analogy which some defenders have adopted. In a democratic voting system each participant has equality as a political right and preferences are counted as having equal weight. This condition does not hold in a free market because people enter the market with different resources: they will differ in ability, in talent, and in material resources such as income and wealth. Hence, some people will enter the market with advantages, others with handicaps, and their preferences will be unequally weighted in the subsequent market transactions. While, of course, individuals bear some responsibility for the talents they possess, nevertheless both for the advantaged and the handi-capped there is a degree, quite a large degree in fact, to which they are being benefited and rewarded or disadvantaged and penalized for factors which are not a matter of their own responsibility. The talents and abilities which people have and to some extent the degree of material resources they possess are the result of factors for which they bear no responsibility. Genetic endowment and fortunate home background within which genetic talent is nurtured are central to the development of personal capacity, including the capacity to act effectively in the market. On the market socialist view these should be compensated for so as to enable people to enter the market on the fairest possible terms.

Of course the neo-liberal will argue that this is a distortion of the true moral position. Genetic endowment and family background are, rather like the weather, matters of good luck or misfortune, not fairness and injustice. Hence there can be no morally justified demand for compensation. In addition, the worst off are not coerced by their lack of resources; they may, of course, do less well out of the market than those better endowed, but their lack of resoures is not a restriction on liberty, because, as we saw earlier, freedom is not being intentionally coerced by another agent.

However, at this point the socialist will want to dispute the account of freedom given by the liberal critic. There will be two aspects to the argument. First of all, the socialist will want to argue that a purely negative theory of liberty cannot fully explain the value of liberty in human life. Why do we want to be free from coercion? Why do we regard it as so valuable not to be subject to another person's will? The answer must surely be that, if we are free from coercion, we are then able to live a life shaped by our own desires and preferences and not those of another, and that is is part of what the distinctively valuable features of human life consist in. However, in order to realize what is valuable about liberty, we have to be able to pursue values of our own, and to do this we have to have abilities, resources, and opportunities—that is to say, some command over resources so that we can live life in our own way. In distinguishing so sharply between freedom and the capacity for agency and its associated resources, the neo-liberal critic of socialism will not be able to explain why liberty is valuable in human life and the conditions which have to exist in order for its value to be realized. Liberals profess to believe in equal liberty, but the socialist will argue that the equal worth of liberty is important too and that this must demand some greater equality in resources for people entering the market.

The second point is that the socialist will dispute the view that markets are not coercive because their outcomes are not intended. The argument here is twofold. The first is that, when Hayek and others deploy this argument, they tend to do so in relation to individuals: that the outcome for a particular individual is neither intended nor foreseen. Of course, this is true, but that has never been the basis of the socialist case, which has been based upon groups and classes. The claim has been that the class of people entering the market with least will derive least benefit and resources from it. This may be unintended, but it can be foreseen, and certainly the experience of the last eight years of Thatcherism confirms it (see, for example, Rentoul, 1987; Walker and Walker, 1987). The socialist will then argue

that we are responsible not merely for the intentional consequences of our actions but also for the foreseeable ones.
Certainly this principle works at the level of personal morality:
if I intend to do Y and I know that X is a foreseeable consequence of doing Y even though the occurrence of X is not part
of my intention, it would be difficult to evade the responsibility
for X. So in a market, if it is a foreseeable consequence of the
operation of a free market with the existing highly unequal
distribution of resources that some will be made poor as the
result of its operations, and if something can be done to change
this—for example through a redistribution of resources—then
it would be difficult to evade responsibility for this outcome
even if it was not part of any individual's intention. In this sense,
if we can link foreseeability and responsibility together, the
market socialist can argue in favour of the redistribution of
resources in order to give to individuals the capacity to act as
effective and free agents in market transactions.

None of this in any way lessens the market socialist's
commitment to markets. Like most social institutions, markets
can be characterized in more than one way. The neo-liberal's
characterization is tendentious and seeks to avoid collective
responsibility for the means which people have at their disposal
when they enter markets.

At the moment it is important to remember that this more
positive view of liberty is not being argued for as an end-state
principle. In fact, it is concerned with the resources which
people should have in order to enter markets in an effective
manner. In this sense it could be said, to borrow a phrase from
Ronald Dworkin, that it is a starting-gate rather than an end-
state principle. It is not arguing that the outcomes of markets
should be adjusted so that people enjoy the same value of liberty
at the end of a set of market transactions if they enter the market
on more equal terms. That is an issue which will be considered
later. It is rather an argument in favour of initial redistribution
so that people enter markets on a more equal basis in terms of
resources.

The liberal critic will, however, argue that even this view of
freedom and markets cannot avoid making highly contentious

moral judgements about what resources people need to be able to enter markets effectively. In this sense positive freedom is a highly moralized conception of freedom and there can be no moral agreement about this in a morally pluralist society. There are three answers to this. In the first place, it is not clear that this is factually correct. While, of course, people do have different moral views, most people in our society would accept that, in order to act effectively, one does need command over certain sorts of resources which define the basic conditions of effective agency and that these will include income and access to education and health care throughout the course of life. Second, if redistribution were in cash rather than in terms of services, then this would avoid many of the problems involved in coming to detailed judgements about needs, with all the difficulties about commensurability pointed out earlier. In addition, if this cash redistribution in the case of health and education were to be given in the form of vouchers, this would avoid the problem pointed out by the neo-liberal critic about the extent of bureaucratic discretion which must accompany the provision of services in kind. There is an argument about freedom here which socialists must take seriously: namely, that we cannot be wholly serious about individual liberty if we completely resist contemplating procedures such as vouchers which would empower individuals against bureaucracies and producer interest groups in the welfare and educational services. These issues are discussed in much more detail in Julian Le Grand's Chapter 8.

The final answer to the neo-liberal's view that the account of freedom given here is tendentiously moralized is to take a leaf out of his book. In his view we have no way of assessing the merits of any particular individual or of the conception of the good held by any individual. These judgements are seen as irredeemably subjective and disputable. If this is accepted, then it could be argued that no individual merits more or less in the distribution of those basic resources which are necessary to enter the market on a fair basis and thus those resources should be distributed as equally as possible because, if the neo-liberal is correct, there is no other criterion which would not involve weighing up incommensurable merits and deserts. Again the equality at stake is not an end-state principle, referring to equality of outcome, but rather greater equality of initial starting-point, institutionalized in ways that will enable people to make their own judgements about health care, education, and the other basic goods of agency. Of course, in the very exercise of these judgements, inequalities will result, but to some degree these inequalities will have to be accepted, partly because, if we respect individual freedom, we have to respect the consequences of the choices which people make and their corresponding responsibility for them. I shall return to this issue later in the chapter.

So far then I have argued that one central plank in the market socialist's case will be the acceptance of markets as procedures for the efficient use of resources and as guarantors of freedom of choice, but at the same time the rejection of the market liberals' characterization of markets as wholly impersonal procedures for the consequences of which we bear no collective responsibility. To secure a really free market we have to be concerned not only with the procedures which a market involves—breaking up monopolies and legislation to ensure that no coercive transactions take place—but also with the conditions of freedom for the individuals who enter markets and with ensuring that these conditions embody in their institutional form the highest degree of freedom of choice. This is not to say that a socialist society should provide no services in kind. Some will always be necessary. But we should not assume that existing forms of service are the only ways in which we can respond to ensure meeting the justified claim that individuals should have command over resources in order to enter markets effectively.

There is a second argument which we need to take seriously about what conditions are necessary to ensure that markets operate in a free and fair way beyond the procedural sense of being governed by a framework of impartial laws. This is concerned with the ownership of capital in a free market economy and the connection between this and the exercise of power in a market. As we have seen, the market liberal wishes to reject theories of social justice—about how resources should be allocated—and he will brook no arguments in favour of redistribution. This bears directly upon the issue of the ownership of capital in a free market. If capital ownership is concentrated, this will enable those who own capital to exercise power over others and will lead to coercive exchanges between those who do and those who do not own capital. This will typically occur in a firm. The workers in a firm who do not own capital will have to work on terms to a degree dictated by the owner of that firm, particularly if capital becomes concentrated and if there are, in a particular community, no realistic work alternatives. This gives the capitalist a considerable degree of power over workers, who will not be like independent subcontractors but rather will be subject to discretionary power, either by the capitalist or by those appointed to oversee his business for him. The neo-liberal will see nothing wrong in such inequality of power and again will not see it as a potential restriction on liberty. This is for two reasons. In the first place, as we have seen, inequalities, however large, in the distribution of material resources are not a restriction of freedom, because freedom and the possession of resources are different things. Only if the capitalist is a strict monopolist will his behaviour be potentially coercive. Otherwise a worker has freedom to work or not to work for a particular firm, and, while this option is open, whatever the position of the worker in terms of resources, he is not coerced by the behaviour of the capitalist. Second, in a free market, capital is accumulated through a process of free exchange. So long as the capitalist does not acquire capital as the result of coercion, then the ownership of capital, however concentrated, is not unjust. It could only be regarded as unjust on the basis of some socialist end-state principle which he rejects. Hence, however concentrated capital may become as the result of free exchange, its ownership is not unjust and the power which it confers is not illegitimate.

However, socialists will be minded to reject both of these arguments. We have already seen the grounds for rejecting the first in the earlier argument about the nature of freedom. The second argument is more complex. One central issue would be whether we lack sufficient historical information to determine

whether present concentrated holdings of capital were justly
(i.e. non-coercively) acquired, and, given the threat to equal
freedom which such concentrations of capital and power pose to
society, a reasonable principle would be to undertake the
dispersal of that capital and property rights in the means of
production more widely in society. This is the view adopted
even by Nozick (1974, 231), one of the arch defenders of free
market capitalism.

There are two ways in which the ownership of capital could
be dispersed: individual and group dispersion. Individual
dispersion would give to individuals some entitlement to the
ownership of capital as a kind of patrimony, perhaps to be
acquired at the age of majority, which would then increase each
individual's effectiveness in the market (as discussed in Chapters
4 and 8). This could be done, either through a negative capital
tax as proposed by Atkinson (1972), or by giving workers share
entitlements after a period of years in a company, so that labour
then created property rights in firms. The other proposal would
be for capital to be owned by the state, which would then lease
capital to worker-owned co-operatives. This latter proposal is
extensively discussed in Chapter 7. The important point to
notice, however, in the light of the arguments discussed so far,
is that again they are both starting-gate rather than end-state
theories: that is to say, they are concerned with ensuring the
conditions of fairer entry into the market and securing those
conditions which will enable markets to operate with the least
degree of coercion. They are concerned with empowering
individuals and groups in the market rather than with criticizing
market outcomes. According to this radical version of market
socialism, the market needs socialism in order to make its
starting-points fairer and more free, but it would neglect the
outcomes of the transactions and exchanges which were then
undertaken in the market.

If individuals and groups such as workers' co-operatives enter
the market on a fair basis, and if the procedures of the market
are fair, then is there any moral basis for criticizing the
outcomes of such exchanges? Traditionally socialists have
wanted to argue that there is a basis to be found in end-state

principles such as social justice, equality, and community. So, for example, it is likely that in relation to workers' co-operatives the end-state socialist argument will go as follows (see also Plant, 1984). The basic problem with decentralized forms of socialism is that, while it may be true that within relatively autonomous decentralized economic units like co-operatives there may well be a high degree of equality of income, power, and status, this does not address the question of the relations between co-operatives and the extent of possible inequalities between them following from their performance in the market. There are perhaps two aspects to this problem. In the first place, in any system of autonomous enterprises, differences are almost bound to arise between such enterprises, because of the differences between internal efficiencies, the skills of workers and managers, accessibility to and relations with suppliers, and consumers, the age and quality of equipment, consumers' choices and demands, decisions about how the earnings of the enterprise are to be allocated between wages, bonuses, services, increasing employment opportunities, depreciation, and invest-ment. In short, the end-state socialist will argue that, without some state-directed redistribution between enterprises, market outcomes are likely to be highly unequal. According to the end-state view, this must be of central importance to socialists; in the same way as the free market has been criticized as being indifferent to distributive outcomes between individuals, so market socialism cannot abandon a concern with equality of outcome between groups.

The issue at stake here is this: does the fact that the market socialist aims to make the market freer and fairer through the reforms which I have discussed mean that the inequalities which will inevitably arise between the workers' co-operatives as the result of their trading in the market are now to be accepted as legitimate, or does the rectification of such inequalities still embody a legitimate moral claim? The issue applies equally to the position of individuals: if individuals are empowered in the market through cash redistribution, are the subsequent in-equalities which will occur as the result of individuals trading and exchanging in the light of their own view of their interests legitimate, or should they be rectified by end-state as well as by starting-gate forms of redistribution? In a sense this is a modern version of the old socialist debate about equality of opportunity and equality of result, or between equity and equality. The radical position on market socialism would be to say that we should be concerned to make the conditions under which markets operate freer and fairer and then accept the inequalities which arise as just and legitimate. A failure to do so will mean forgoing most of the advantages of the market: its competitiveness and dynamism, its capacity for innovation and change. If people know in advance that there will be equality of result however they act in the market, this will be a recipe for inefficiency. In addition, the old problems about end-state socialism will re-emerge: how do we produce some consensus about the degree and nature of equality? End-state socialism implies a large and necessarily bureaucratic state together with a commitment to detailed planning to make sure that market outcomes will conform to the desired form of end-state equality. So in this sense market socialism would require a radical revision of traditional socialist understandings of equality and social justice.

The same is true of community. This applies in two ways. As I have argued, a radical market socialist view of the role of welfare might well favour a voucher system, for example, in the spheres of health and education, to empower people against bureaucracies and give them real choice about the type of health care they want and the sort of education they desire for their children. However, it is obvious that this poses a major challenge to one of the traditional socialist justifications for the service rather than the cash or voucher provision of welfare: namely that, if people undergo the same experiences in school and in health, this will foster a sense of community and common culture. This would be lost in a cash or a voucher system. Take a sharp real case as an example. In the educational sphere one could imagine that this might lead to a development of ethnically based schools. For example, Muslim families might use their vouchers to send their children to schools which are based on Islamic precepts, or, at the other end of the spectrum, white families might choose to send their children to schools which had few, if any, children from ethnic minorities. In these ways, far from an educational system developing a sense of common culture and community, a voucher system might well become a vehicle for ethnic absolutism of all sorts. According to an end-state socialist view, the state has a duty to foster as far as it can a sense of common culture and community and must use regulation and zoning to achieve this. In this sense bureaucratic procedures have to be used to produce a particular end state, in this case a greater sense of community. The radical market socialist might argue that the sense of community involved here is a delusion, because community is just not possible in this overall sense in a modern complex, individualistic society; what is important is that people should have the choice to contract into those forms of community which seem to be important to them—perhaps, in the case we are considering, an ethnic community.

If socialism is to be allied to increasing liberty and freedom of choice, it should not seek to impose a particular pattern of community on society, but rather accept the diversity of community forms which will emerge as the result of people exercising their own choice. You cannot on the one hand seek to empower people and then restrict in an artificial way the choices open to them in pursuit of some ill-defined concept of community. The only restriction which could legitimately be placed upon freedom of choice would be to restrict the exercise of choice which limited the capacity of other people to exercise their choices in an effective and meaningful way, and it is not clear that in the schooling example we have been considering this would happen. Hence the situation is precisely similar to the case of equality: the end-state socialist wishes to use government to produce a particular social pattern whether of greater equality of outcome or of a greater sense of community. The radical market socialist's position is to argue that the important thing is to give people effective power in a procedurally and substantively fair market and then accept the end results of that process as legitimate. Freedom and power for individuals and groups means that we have to accept the results and not reject them because they do not conform to a chosen pattern. Hence it appears that market socialism in a thoroughgoing form places a major question mark against traditional socialist assumptions.

TWO VERSIONS OF SOCIALISM OR ONE?

In the final part of this chapter I want to consider the extent to which a reconciliation between these two apparently different forms of socialism is possible. What is at stake here is the nature and role of government in a feasible market socialist society. Much of the impetus for market socialism has come from a sense of disillusionment with statist forms of socialism (see, for example, Nolan and Paine, 1986). Obviously it is difficult to see how an end-state form of socialism could operate without an extensive role for government—creating the framework of law within which market forces would operate, intervening in the market to secure outcomes consistent with socialist values, and providing a non-state sector of welfare services in order to remedy market failures in this area. A radical form of market socialism, however, shares with the neo-liberal the not implausible view that in modern economies we have been confronted by government failure as much as market failure and many of the assumptions of market socialism are based on alternative forms of provision to centralized state action.

My own view is that some end-state conceptions are inherent in any plausible form of socialism and indeed enter into the various market socialist proposals which I have been discussing. If this is so, then, at least in this respect, the debate is not market socialism versus state socialism, but rather an explicit endorsement of a central role for markets in a socialist economy (which has not always been forthcoming within the socialist tradition) within a framework both legal and substantive set by government guided by end-state or patterned principles. The neo-liberal project of procedural justice cannot be made fully compatible with socialist ends. Socialism does require certain kinds of outcomes, not just those which emerge as the result of fair procedures. There are two important issues here: first, to argue the case for markets within a socialist economy but,

second, to do so in a way which makes them compatible with the
maintenance of patterned or end-state principles such as greater
equality and social justice. It seems clear from the examples I
have given that market socialists do require end-state values to
underpin what might at first sight appear to be procedural
recommendations. Take, for instance, the issue of redistribution
in cash in order to empower people to be able to enter the
market on a more equal basis. The degree of such redistribution
if it is not to be arbitrary is going to have to be grounded in
some patterned notions—most obviously need, effective liberty,
and social justice. That is to say, the degree of transfer between
the better off will have to be guided by some conception of what
needs have to be satisfied in order to secure for individuals the
capacity for effective agency in the market. This judgement
cannot merely be left to people's revealed preferences to avoid
the problem of some political judgement about need, because
the whole argument assumes in the first place that people's
revealed preferences will differ in relation to the initial resources
which people have: the rich are likely to have more extensive
revealed preferences than the poor. Hence we cannot just look
to the demands which people actually make to determine what
they need. The issue here was posed clearly by Runciman
(1966). The worst-off members of the society have very limited
preferences, because they compare what is possible for them not
with the rich but with those only slightly further up the social
scale. Runciman concludes, rightly in my view, that a theory of
justice is absolutely vital in order to determine what needs
actually are: what constitutes a legitimate and an illegitimate
claim on resources. Social justice will enter in another way too,
namely in trying to work out what should be distributed
according to need and what should be left over for market
rewards. This is a point made very clearly by Miller and Estrin
in their contribution to the Fabian symposium on market
socialism when they argue, 'It is quite feasible to think of a
division of social resources between those earmarked to satisfy
needs and those serving to reward merit, and to provide the
incentives necessary to make a market sector function effectively'
(Forbes, 1987, 11). This is clearly correct, but, as Miller and

Estrin point out, this does require the development of a theory
of distributive justice: that is, a patterned or end-state principle
which, as we saw, the neo-liberal argument cannot countenance.
The same is true of other aspects of the market socialist case. For
example, in the context of a market socialist case for vouchers as
developed by Julian Le Grand in Chapter 8, considerations of
equality enter in a very central way. Such a voucher system goes
far beyond a starting-gate theory and involves central considerations about equality which again is a patterned principle
which it would be the duty of government to implement. Again
considerations about community would also enter in Le Grand's
reference to the likelihood of wanting to have a national
curriculum rather than one which was geared wholly to the
prescriptions of a particular set of parents about what their
children should be taught. The same holds true for the
argument about the redistribution of concentrations of capital
which are likely to occur in a free market system, which was
discussed earlier in the context of dispersing capital to workers'
co-operatives. This could only be done in a principled way with
reference to a developed and patterned principle of social justice.
This would apply in two ways: first to provide the basis for
criticizing the concentration of capital, particularly if it could be
demonstrated that capital was acquired through uncoerced
exchanges; and, second, to guide the degree of redistribution
and dispersal of capital holdings in the economy.

CONCLUSION

All of these observations lead to two conclusions. First, even
market socialism needs a theory of distributive justice, equality,
and community, and this means that market socialism is a long
way from merely humanizing and making more fair the neo-
liberal project in relation to markets. Second, the maintenance
of these patterned principles must presage a central role for the
state, and market socialism cannot be seen as a panacea for the
problems of government. The central issues facing socialists in
this context are, therefore, twofold. One is to argue the case for
markets and explore forms of market provision in ways which

may well upset many traditional socialist assumptions, as well as the producer interest groups who have a vested interest in maintaining those assumptions. The second is not to be seduced by those siren voices which assume that an advance towards socialism can be achieved without a powerful state. As socialists in Britain, we need to develop a theory about the role of the state as much as markets and to meet the neo-liberal challenge to those patterned principles of justice and community without which socialism will not be a viable intellectual or practical project. In the same way as market socialist views about the role of market have been concerned with empowering individuals and groups in the market, the corresponding impetus in relation to the theory of the state should be the empowerment of individuals through extending democracy and accountability in both political structures and bureaucracies.

4

An Equitarian Market Socialism

Peter Abell

If socialists are to be absolved of the accusation that they speak of lands which cannot be inhabited without an unwarrantable surrender of human liberties, then they need to show how satisfactory and mutually compatible conceptions of efficiency (productive and allocative), freedom (positive and negative), and justice can be woven into a single garment. If each could be gained without the surrender of the other (an assumption of most socialist theory), then it would merely be a matter of showing how. But in general they cannot—even if we postulate fundamental changes in human values—and, thus, ways of balancing the possible trade-off between them need to be constructed.

The terms of this trade-off will be dependent partly upon unalterable factors (such as the distribution of innate ability[1]) and partly upon changing and changeable values, beliefs, and motives which the people bring to both production and consumption. The challenge for contemporary socialists is to formulate a normative model of society which can serve as a template against which proposed policies may be assessed. This model must be libertarian in spirit, both embracing and transcending the liberal concept of negative freedom. It must also be sensitive to the understandings which contemporary social science affords us. It must not be over-sanguine about the

---

[1] I will assume throughout that innate abilities are not equally distributed—or, at least, that those abilities which are supplied in the relations of production are not. I realize some will jib at this. My conclusions, however, would not be altered if such abilities were identical in all respects; indeed they could be reached that much more easily if one were to assume equality. So to that degree. the assumption is innocuous.

malleability of human motives and institutions but, at the same time, it must not be over-pessimistic, leaving space for the better side of humankind eventually to occupy.

Humankind is probably not perfectable, but surely it can be coaxed into creating something better than we now have.[2] It is futile and dangerous, as Anthony Crosland observed, to advocate an ideal society tomorrow, but having some model in mind, guiding and informing a permanent transition, is another matter. This is so even if, in some sense, the model is ultimately unattainable in its extreme form (such as perfect competition). Such models, which we may describe as regulative, offer a vocabulary within which the issues at stake may be rationally discussed and whereby reasons may be provided for the unavailability of the extreme form. Equitarian market socialism should be seen in this light.

I shall argue that the essence of the socialist vision rests with a progressive attention to the satisfaction of human needs through the equalization of human agency (that is, positive freedoms within the framework of negative freedoms) primarily in production and only secondarily in consumption. Socialism is about the eradication of poverty and is about greater equality of opportunity; but it is about these things because it is about greater equality of freedoms.

It is an unfortunate feature of our time that the moral highground in respect of issues about choice and freedom has been effectively commandeered by the New Right. The Left is frequently, and not without justification, castigated for its advocacy of large bureaucratic structures which inevitably fail to respond to people's needs and restrict their choices. This is equally true of welfare, education, and economic institutions. The identification of socialism with both bureaucratic sloth and

---

[2] It has become unfashionable to speak of building institutions around changed values. Certainly caution must be exercised; it is all too easy to assume away problems by relying upon unrealistic values which are more congenial to a socialist ethic. Nove (1983) has offered us a cautionary tale in this respect, and premature attempts at institution building, in China for instance, underscore Nove's caution. Nevertheless socialism is about the interplay of institutions and values, and it would in my view be equally wrong to reify existing values.

restrictions upon choice has, to a very considerable degree, arisen as a consequence of the Left's distrust of markets (see Chapter 2) in favour of the planned allocation of resources (capital and labour) and of goods and services.

This distrust is not without foundation; markets can be both inefficient and unfair, failing to take account of all costs and benefits and generating great inequalities in income and concentrations of wealth. But it is important to distinguish between the unjustifiable consequences of markets *per se* and those consequences which arise as a result of the different endowments which people bring to production.

It is salutary to note at the outset that the inequalities which markets almost invariably generate are attributable to one or more of three factors—namely, unequal endowments in production, lack of competitive conditions, and inescapable market uncertainties. To put it another way, in that perhaps ultimately unattainable world, with perfectly equal productive endowments (including capital, labour skills, information, etc.) and with certain and perfectly competitive markets, all incomes would be equalized.

The market socialist wishes, where possible, to reap both the efficiency and libertarian characteristics of markets whilst promoting much greater equality than we presently experience. The market socialist society will inevitably require a strong democratic state, however, with powers to intervene and regulate where markets fail, with powers to promote competitive conditions and undermine monopolies, and, above all, with powers to promote equality of freedom.

It should perhaps be emphasized here that in what follows equality (of both positive and negative freedoms) is construed as a value in its own right. Thus, I will argue that maximizing human freedoms subject to an equality constraint is constitutive of socialism rather than the more usual, social ownership of productive assets. If the latter leads to the former, then fine; if not, then there is no reason to endorse it. Some socialists may, however, prefer a more elaborate moral framework, showing how such freedoms might be used. This framework falls beyond the scope of this chapter.

IN SEARCH OF PRINCIPLES

The clarion call for an earlier generation of socialists was uncompromising and clear. It was to adjust socio-economic arrangements, in the first instance, when humankind is still imbued with the dispositions of the pre-existing acquisitive society, so that they embody the principle (derived from Marx, 1875):

(P1) 'From each according to ability, to each according to ability.'

And later (or progressively), when these dispositions (assumed malleable) have dwindled, to readjust to the principle:

(P2) 'From each according to ability, to each according to need.'

Both of these principles capture in an appealingly pithy manner ideas of efficiency ('from each according to ability') and distributive justice ('to each according to . . .').[3]

But do they still carry conviction now that we have the benefit of hindsight in respect of the 'socialist' societies (and islands of 'socialism' in mixed economies) and now that we are also better equipped, in virtue of the development of the social sciences, to understand the complexities of human institutions? I shall argue that when taken literally they no longer do, but, nevertheless, when suitably buttressed they can still serve as a useful guide; not a guide, though, that enables us to duck some rather hard decisions.

There is no mention in either principle, for instance, of matters concerning liberty. P2 is also silent upon how incentives are to be structured in order to guarantee the compatibility of its two constituent parts, whereas P1 assumes a particular structure

---

[3] This implication is of course hedged by a *ceteris paribus* condition. In order that we might guarantee the full utilization of ability, we would also need to say something about the distribution of other factors of production. Furthermore, the phrase might be more correctly read to imply the optimal provision of ability within the framework of these other factors. On either reading, full employment is presumably also implied. For Marx, the juxtaposition of 'from' and 'to' ability follows from his assumptions about the labour theory of value and consequent exploitation in the capital–labour contract. P1 is thus a plea for the eradication of exploitation.

of incentives: namely, that it is necessary to pay the able more to encourage them to use their ability. We must therefore search for more secure foundations upon which to fashion a contemporary socialism.

Principles P1 and P2 both contain the same initial phrase 'from each according to ability', implying that, if human abilities are afforded full sway in productive relationships, then this should maximize the value of goods and services delivered for eventual consumption. It is, thus, a putative principle of productive (and implicitly allocative) efficiency. It invites us to find a way of arranging production such that human abilities may be best utilized (i.e., in Marxian terms, so that at least one of the forces of production is not fettered). If this was all that was implied, then the phrase would not be particularly controversial; but it may, in addition, be argued that the phrase also derives from a deeper assumption about the 'dignity of labour'—that is to say, an assumption whereby the relations of production should be so constructed as to allow for the maximum feasible expression of ability. To put it another way, people's 'needs' should, where possible, be addressed by making way for their effective agency in production.

One of the prime objectives of socialists has always been to reduce or eliminate alienation; giving full vent to human abilities must surely be one essential aspect of this. But, in addition, I am making a slightly stronger point: that, given a feasible choice, then it is usually preferable to satisfy human needs by enhancing their productive capabilities rather than by merely increasing their income.

I thus construe the phrase 'to each according to need', as a matter not merely of matching consumption to needs, but also, where possible, of enhancing people's capability of production, so that they may themselves, satisfy their needs. My argument will be that this is best achieved within the framework of an equitarian market socialist economy.

In the context of both principles P1 and P2, one may, of course, distinguish between innate (or natural) and acquired abilities (or acquired human capital). If one does so, then it is not clear which type of ability the principles P1 and P2 refer to.

Moreover, if it is acquired ability, then what proportion of current resources should be employed in acquiring abilities and how should they be distributed? In addition, how should the other factors of production—notably capital—be allocated to match the supply of ability (innate plus acquired) in order that efficiency may be maximized? Answers to these questions solely from an efficiency standpoint are relatively straightforward but do depend upon the assumptions we care to make about the relationship between expenditures and the acquisition of abilities.

But this is not the end of things. From a socialist perspective, both the enhancement of individual capabilities and the opportunities afforded by the 'possession' of productive capital (socialized or private) impinge not only upon matters of efficiency but also upon positive liberty. Are not better educated and endowed individuals capable of a wider range of choices (both as producers and consumers) and are they not, as a consequence, in some sense more free than they otherwise would be? If this is granted, then, when seeking to expend resources on the acquisition of abilities and in allocating capital to people, we may properly be challenged to justify our practices on grounds of both efficiency and liberty. And, unless we are very lucky, these objectives may collide with each other.

Thinking of liberty in this way is not, however, uncontroversial. The prevailing liberal orthodoxy would tell us not to, and, by so doing, conveniently manage to avoid a possible collision. People, it is averred, are free to the degree that their potential actions (whether or not they have the wherewithal to realize them) are not intentionally and foreseeably impeded by others. This essentially negative concept of freedom should, furthermore, be equally available to all and at a maximum volume compatible with this equality (Rawls, 1972). If one endorses this principle, then in practice it amounts to little more than an acceptance that all should be equally placed before minimally restrictive laws. There is, as far as I can see, no reason for a socialist to reject this conclusion about negative freedoms unless it runs counter to other objectives.

The *exclusive* concentration upon the specifically negative aspects of the concept is, none the less, unacceptable. Are we to regard the fabulously wealthy and well-educated individual who is identically placed before liberal laws alongside an impoverished illiterate as equally free? Surely there is a manifest perversity in so doing? As discussed in Chapter 3, freedom should be defined so as to embrace not only the absence of coercion by others but also the possession of those resources which afford the grounds upon which effective agency and choice are based. Freedom has a negative and a positive face.

But should the possession of all resources enter into the definition of a positive conception of liberty? Some, dissatisfied with the complete exclusion of a positive face, have spoken of 'basic resources' which, they argue, are necessary to capability or effective agency. Only these, it is urged, should enter into a definition of liberty, not the wider set of resources. But how are we to arrive at such a conception—at the appropriate definition or specification of 'basic resources'? I can see no possibility of doing so; surely any such conception would always be arbitrary.

Inevitably, it seems, we have to go one way or the other; either we follow the liberal orthodoxy and retreat into a sole reliance upon negative freedom, or we accept the full implications of endorsing the view that all resources, in one way or another, provide grounds for effective agency and thus may properly be said to impinge upon an individual's positive freedoms. There is no half-way house. If the effective use of resources carries undesirable consequences (people are free to do rotten things for instance), then presumably attention to the negative face should handle these eventualities.

Clearly the central problem we face is that the distribution of resources (i.e. alienable ones like capital and income, and inalienable ones like ability, skill, etc.[4]) carries implications for both efficiency and positive freedoms. Can we find a way of efficiently bringing forth people's abilities ($P1$ and $P2$) whilst at the same time satisfying any equality constraints we might wish to impose upon the ways in which positive freedoms are shared?

Issues of efficiency are not terribly controversial. Under fairly reasonable suppositions about how expenditures transmute into

---

[4] We may construe acquired abilities as inalienable once acquired.

acquired abilities, one would, in seeking to promote productive efficiency, spend more upon the able.[5] Furthermore, allocative efficiency would dictate that alienable factors of production should also be disproportionately placed at the disposal of the more productive (i.e. able).[6] The conclusion on both counts is thoroughly inegalitarian in favour of the more gifted. One would expect that, given distributions of ability (innate plus acquired) and alienable factors, then efficiency would be maximized in production and everyone would be as well off in consumption as they could be if P1 held sway.

If efficiency were to be our only objective, life would be comparatively simple. Presumably we would be searching for the best institutional set-up which would procure these efficiency characteristics. Of course, incentives aside, P2 could also be satisfied by redistributing income in consumption in order to satisfy identifiable needs. Traditional socialist theory has sought to promote these objectives through the planned allocation of resources underpinned by welfare provisions. It is at least open to doubt whether such arrangements do in fact achieve

[5] Learning theory would lead one to postulate logistic individual learning curves relating ability to resources expended. If we assume that rational calculations concerning the delivery of ability cover the concave range of such curves (i.e. positive but declining marginal acquired ability per unit of expenditure), and that those of greater natural ability have steeper learning curves, then the conclusion in the text follows.

Furthermore, if the productivity of the economy depends not only upon the total supply of ability but disproportionately upon the mix of higher abilities, then the unequal expenditure implications would be further supported. If all individuals had identically shaped learning curves (differing only in the intercepts upon the ability axis), then equal expenditure would maximize the supply of ability but not equalize abilities, of course. It would be helpful if we could make this assumption, but there are no good reasons to do so. What is more, the mix-of-abilities argument might still hold.

The assumption that expenditures might bring us into the concave regions of the learning curve might cause some readers concern. If the curves are increasing linear, then maximizing total supply of ability per unit of resource would allow expenditures to be applied in any way from egalitarian to all on one. If the curves are convex, then extreme inegalitarian conclusions follow.

[6] This statement needs careful interpretation. It may be read as the usual allocative principle concerning marginal productivities and incomes (P1). However, nothing follows about the income generated until assumptions about incentives are made. If there were no incentive problem, then of course the factors determining maximal efficiency in production and those determining the distribution of consumption could be fixed independently.

efficiency, but this is not the point at issue here. For all of this would be to ignore what I have suggested is the socialist ideal of, where possible, satisfying needs by equalizing positive freedoms in production rather than in consumption.

So, let us now assume the objective is to equalize such freedoms. Here we are unfortunately on much less secure ground, as there is little or no social scientific understanding of how 'abilities' and the 'possession of alienable assets' translate into such freedoms (i.e. into ranges of effective choice or agency within the framework of equally available negative freedoms). Under, however, what seem to be fairly reasonable suppositions (though I admit not entirely uncontroversial ones), it does appear that conclusions are much more in favour of the needy (i.e. less able). To wit—that the degree to which (1) educational expenditures are designed to reduce assumed unequal innate abilities by disproportionate expenditures on the less able, and (2) alienable productive assets are equally distributed, then positive freedoms are equalized in production. And also (*in extremis*) conveniently maximized.[7]

This rather technical sounding conclusion is merely saying that, under the reasonable assumptions outlined in note 7, if we wished, at any instant, to equalize and maximize positive freedom solely in the relations of production, then this would be accomplished to the degree that differences in innate abilities are ironed out by helping the less able and by giving everybody an

---

[7] Assume that all individuals have identical freedom functions in ability: that is to say, other things equal, people of the same ability are equally positively free. So positive freedom functions in ability are 'objective'. Positive freedom is thus not to be equated with the value of utility or freedom (cf. Rawls, 1971). This assumption may be counterfactual in respect of some individuals at higher ranges of ability. This issue aside, if the function is concave on to ability, then the conclusion in the text follows. If it is increasing linear, then one way of maximizing positive freedoms is to equalize ability. A convex function would suggest lavishing expenditures on one randomly chosen individual.

Similar reasoning may be applied to the relationship between positive freedom and alienable resources in production and income in consumption. Depending on assumptions about the elasticity of substitution between alienable and human capital, it may be possible to compensate those with a limited capacity to acquire abilities with alienable resources. The positive power function in alienable assets may perhaps be linear (cf. Rawls: 'the worth of liberty is proportional to their capacity to advance their ends . . .').

equal access to the other factors of production. Nobody of course believes that this is at all practicable when taken to extremes, but it does seem that the venerable socialist adage about equalizing access to the means of production is entirely consistent with the objective of equalizing positive freedoms in production.

Expenditures and distributions of the sort described get in the way of efficiency; however, what is more, in reducing the total available income, they would not maximize positive freedoms (i.e. choices) in consumption either. But of course, in the context of perfectly certain and competitive markets and identical preferences, they would equalize incomes which, in turn, would equalize positive freedoms in consumption; but only if all needs had been addressed at the point of production.

Furthermore, by reducing the total income, this would reduce the funds available for reinvestment and growth and thus ultimately the positive freedoms in production also. This would come about for three reasons: first, because of the suboptimal pattern of expenditures upon the acquisition of abilities; second, because of the suboptimal allocation of capital (assuming all cannot in the final analysis be brought to the same level of ability), and, third, because of the disincentive effects implied by the redistribution necessary to maintain the equality of positive freedoms in production.

It is still not evident, however, whether the strong egalitarian implications should apply at one point in time (e.g. on entering the labour market) or over a lifetime. On pragmatic grounds we must, I think, endorse the former procedure, underpinned by welfare provision for those who fall by the wayside for one reason or another. Apart from anything else, guaranteeing lifetime equality smacks of excessive bureaucratic intervention, something we are trying to avoid. So I am suggesting that socialist policies should be directed towards a radical perspective upon equality of opportunity—a perspective ethically grounded in terms of positive freedom.

One would not expect perfect initial equalization—even if one could compensate those of less ability with more capital. Nor would one expect to eradicate all market uncertainties. This being the case, income differentials will emerge and intra-generational accumulations of wealth also (if permitted). These would be very modest though—certainly compared with what we are currently used to.

If, in pursuit of greater initial equality, current income is taxed away or alienable factors of production are heavily redistributed through inheritance taxes and so on, then there may be a significant impact upon savings and work incentives. Marx was of course fully aware of this problem and proposed that P2 could only become the operative principle: '. . . after labour has become not only a means of life but life's prime want; after the productive forces have also increased with the all round development of the individual, and the springs of co-operative wealth flow more abundantly—only then can the narrow horizons of bourgeois right [i.e. P1] be crossed in its entirety . . .'

Here Marx is clearly writing about changed human values, or, as he puts it 'the all round development of the individual', and, of course, greater abundance also. Taking a radical approach to the equalization of positive freedoms is clearly easier to the degree that individuals are prepared to accept the sacrifices implied by the needs principle.

It has unfortunately become common practice to consider the incentives issue as fixed by the parameters of 'human nature'—a nature which is in most significant respects egoistic. All the social science evidence tells against such a simple view, however. People, perhaps increasingly, when they feel secure, are capable of altruism and can become committed to wider social groupings—the family, the community, and even beyond.

A contemporary socialism must be about our understanding of changing and changeable human values. It must also be about the quest for those institutions which, whilst not perhaps relying upon altruism, nevertheless make space for it. Clearly the objective must be to encourage those values which permit the progressive introduction of the needs principle. To the degree that values license motives which do not get in the way of efficiency alongside redistributive and fiscal policies which help the needy, then we have achieved a socialist ethic. To the degree that a socialist ethic pervades the people, then one can that much more easily pursue the objectives of a normative model of socialism.

So, what are the principles which should animate a socialist society? As I argued earlier, it would be foolish to seek a set of binding principles. But I have urged that the essence of the socialist vision rests upon the satisfaction of human needs through the equalization of positive freedoms (within the framework of equally and maximally available, compatible negative freedoms) in the relations of productions. Such an objective will normally have an adverse effect upon total income (i.e. total positive freedoms in consumption); as a consequence it can only be progressively pursued as values are eased into an acceptance of the implied redistribution effects and as society gains in wealth. In so far as needs cannot be addressed in this manner (health, etc.), then equalization of positive freedoms in consumption will prove appropriate.

But which institutions will best secure our basic objectives of progressively equalizing positive freedoms in production?

POSITIVE FREEDOMS AND CLAUSE 4

How are we to achieve a socio-economic system which begins to embody the rather abstract characteristics outlined above? Given its historical significance to the labour movement, it is useful to start with the celebrated Clause 4 of the constitution of the Labour Party, for it captures in many people's minds the essence of traditional socialist thought:

> the party is to secure for the workers, by hand and by brain, the full fruits of their industry and the most equitable distribution thereof that may be possible upon the basis of the common ownership of the means of production distribution and exchange and the best obtainable system of popular administration and control of each industry or service.

But can this clause be read as following from our earlier stated principles? The wording of the clause affords pride of place to consideration of property rights over matters of distributive

justice. It is not the most equitable distribution which the
Labour Party is exhorted to seek, but the most equitable
distribution based upon the common ownership of the means of
production and exchange. Why should this be so? Is it because
common ownership is deemed to carry its own intrinsic worth
or is it because there is a supposed empirical relationship
between equity and the disposition of property rights implied?

The common ownership of the means of production and
exchange is by no means a transparent phrase and any answers
which might be proffered to these fundamental queries will
doubtless depend upon the interpretation it is afforded. The
most common one, of course, is nationalization (or municip-
alization), where the full panoply of property rights implied by
ownership is handed over to the state or municipality. Certainly,
the Labour Left has almost invariably been seized by this view.

Intellectual support for a traditional reading of Clause 4
arrives from two interrelated directions. First, there is a belief
that, by taking the instruments of production into common
ownership (for which read nationalization), their use can be
planned in a way which more effectively satisfies initially P1 and
eventually P2. Second, there is hostility to the private ownership
of the means of production—hostility deriving from a deeply
entrenched rejection of the legitimacy of profits accruing to
capital and thus the profit motive. The premiss from which this
hostility flows is largely Marxian in inspiration, that is to say,
the labour theory of value—a doctrine which, though widely
rejected outside Marxian circles, still grips the Left and finds the
value of commodities entirely attributable to the quantity of
labour required to produce them.

Armed with the labour theory of value, Marx and those who
choose to follow him believe they are able to demonstrate that
the putatively voluntary market contract whereby capital hires
labour is systematically exploitative and unjust. Furthermore,
the capital–labour contract sets the two protagonists irreconcil-
ably against each other along class lines and institutionalizes a
seismic diversity of interests.

Nationalization of the means of production, it is averred, will
at least bring any profit into the publicly administered domain

where it can be used for the benefit of all (e.g. to satisfy needs). So, nationalization apparently becomes necessary for both the efficiency and the justice of the socialist society. No wonder Clause 4 has caused all the commotion it has. But what are we to make of all this in the light of twentieth-century experience? Can we happily advocate nationalization, planning, and the destruction of markets in the belief that this will cultivate the sorts of equality in positive and negative freedoms advocated above?

Let us start with the labour theory of value and inter-class exploitation. The skein of the above arguments would be decisively severed if the labour theory of value were to fail. John Roemer (1982), building on earlier authors, has, with his celebrated class-exploitation correspondence principle, formally demonstrated that there is only a direct relationship between a person's class position (as worker or non-worker) and his or her exploitation status (as net contributor or recipient of labour time) under entirely unrealistic assumptions. So, even if we are inclined to endorse the labour theory of value, the picture it portrays of class divisions within society are hopelessly diffused and certainly not predictive of socio-economic change.

In addition to these rather technical considerations, many other factors (e.g. social mobility, pension schemes, share options, etc.) have combined to render the line between capital and labour profoundly unclear.

These considerations make it meaningless to postulate an unbridgeable gulf between capital and labour and to adumbrate theories of inter-class conflict, socio-economic change, and distributive justice articulated exclusively around this axis. This is not to say, of course, that there are not vested interests in maintaining and reducing the inequalities of power, wealth, and income; the problem of distributive justice remains, but it must be addressed in a manner which betokens the individual's involvement in both capital and labour and, indeed, as a consumer also.

From this perspective, it is not capital or profit which is 'bad' but its inequitable distribution and the way in which capital arrogates exclusive property rights unto itself. Indeed, it does seem that the contemporary socialist should be ethically neutral between equal access to productive assets which are socially owned and those owned privately but equally distributed. The choice between one or the other is primarily a matter of efficiency (allocative and productive), not a matter of distributive justice.

The fault with Clause 4 resides in the fact that it attempts to solve both the problem of distributive justice and that of economic efficiency at the same time. These two objectives should be addressed separately—the case for nationalization resting upon whether or not it can be proven to be more efficient or, in the case of technical monopolies, more easily regulated.

A NEW FRAMEWORK: AN EQUITARIAN MARKET SOCIALISM

A new framework which would put a rather altered construction upon Clause 4 consonant with our deliberations about positive freedoms could establish those on the Left as the champions of individual liberty and as the enemies of bureaucratic tyranny and sloth. The framework would run somewhat as follows:

1. that 'common ownership' be taken to imply the most feasible approximation (given current values etc.) to an equal distribution of positive freedoms in the relations of production;
2. to the degree that an 'equitable distribution' is not addressed by (1), it be procured by some equalizing of positive freedoms in consumption (welfare provision); and
3. that 'popular administration' be taken to imply democratic procedures which take cognisance of individuals' interests as providers or labour and capital (and as consumers also, if efficiency dictates socialized assets of, say, technical monopolies).

Let us now sketch in a little more detail what this construction of Clause 4 might imply.

First, full recognition must be given to the observation that each individual potentially possesses interests in each of three fundamental socio-economic roles—as provider of labour (i.e. ability), as provider of capital, and as consumer. Structuring socio-economic institutions exclusively around the interests inherent in any one role-type is unacceptable.

A not unjustified criticism of the conventional interpretation of Clause 4 has been that 'socialism' is concerned to assert the rights of individuals as providers of labour, particularly at the expense of their rights as consumers. Moreover, their rights in terms of access to productive capital were usually assumed to follow from the introduction of nationalized assets.

It was perhaps not unnatural that such a view should have come to prevail in a period when there was a sharp divide in society between those who owned capital and those who hired their labour (i.e. a class divide in Marxian terms). The accumulated power of capital and its concentration in a few hands positively invited this response. And solutions to the problems of distributive justice were inevitably bound up with assertions about the rights and entitlements of labour. It is of course open to debate whether 'nationalization' does on balance benefit labour; be this as it may, the market socialist seeks to balance the claims of the provider of labour, the provider of capital, and the consumer, broadly speaking, through the agency of competitive markets. But these should be markets where equality of positive freedoms in all three capacities is progressively addressed.

So, second, economic transactions should, in recognition of their indisputable efficiency characteristics, normally follow the dictates of competitive markets. The state should encourage competition and discourage monopolies of whatever sort. The socialist is opposed to monopolies for they, in effect, reduce both negative and positive freedoms. To the degree that factors of production are equalized, competitive conditions encouraged, and uncertainties reduced through indicative planning (see Chapter 5), then incomes will also be equalized. If the movement of resources from low-earning to high-earning activities is costly, then the state may also subsidize such movements.

Third, as a provider of labour, equality of positive freedoms should be encouraged by:

1. appropriately directed expenditures upon education training and so on;
2. participation at work on the basis of one person one vote (i.e. various forms of producer democracy, see below);
3. the state supporting job creation and providing the environment for new ventures where market forces prove too sluggish; and
4. a reduction in property rights of capital consonant with concessions to producer democracy (also see below).

Fourth, policies should be fashioned to maintain as far as feasible equality in the initial endowments of 'equity capital' (which would, though, carry reduced property rights consonant with producer democracy).

Each individual would bring to the relations of production approximately equal endowments. This should not be read to imply, of course, that they would necessarily work the capital themselves but rather that investments would be made with the perspective of an appropriate income stream. Various investment trusts (controlled democratically) designed to spread risk would become necessary and it would be the responsibility of the state to promote and regulate these.

The introduction of equal positive freedoms will no doubt prove progressively more difficult to achieve as the implied inheritance tax begins to bite and, given current values, the disincentive to save and create new capital, particularly in later life, would be strong. Taxation on consumption may have some effects in offsetting this tendency and the introduction of a capital receipts tax rather than a conventional inheritance tax will help spread wealth voluntarily (see Chapter 8). Indeed, the state should do everything in its power progressively to work towards an equalization of initial capital endowments; there may even be a case for donating some capital stock to all at the age of majority in the form of unit trusts of limited redeemability (again, see the discussion in Chapter 8). With a more capital intensive future, then income from capital initially provided by the state to individuals may well appear as natural as unemployment benefit does now.

For well-established technical reasons (see Chapter 7), relying upon a co-operative economy, where members both own and manage their enterprises (such as the Mondragon co-operatives in Spain) is not likely. The concentration of risk and implied suboptimal level of risk-taking is too high; and the inflexibilities of tying capital and labour together in this manner is ultimately inefficient. Moreover, labour-managed enterprises may under certain assumptions about their objectives 'under-employ' at least in the short run. Labour co-operatives may be most appropriate in the labour-intensive low-risk sectors and should be actively encouraged, where appropriate, with the hope that they might reap the benefits of motivational and organizational efficiencies.

Capital intensive and risky sectors are another matter. Here various forms of labour–capital partnership should allow for variable divisions between capital and labour of control, of risk-taking, and of share in revenues.

Indeed, we may envisage a range of enterprise types running from, at one pole, the pure labour co-operative, through debt-type co-operatives, through equity-type co-operatives, to various forms of labour–capital partnership. Note that the selection of the most appropriate type (given capital requirements, disposition of risks, etc.) is purely a matter of efficiency, not of justice, as the just access to capital and human capital will have been addressed elsewhere.

As Estrin observes (Chapter 7), with an increased emphasis upon markets, as opposed to plan, then the particular form enterprises adopt takes on an added importance. In my view, given the disposition of capital implied by attempts to equalize positive freedoms in production, the stigma attached within traditional socialist theory to both the control by and returns to capital loses much of its sting.

A more relaxed approach can be taken to finding some suitable way of combining the interests of capital and labour within and without the enterprise. Putting the problem more generally, we are searching for arrangements which have the following characteristics:

1. The incentives should be such that labour and capital, where possible, find common interests in conducting the affairs of the enterprise, in particular, in sharing risk (see Estrin, Chapter 7) and fixing the levels of labour and capital employed. The assumption here is that such arrangements will enhance the motives of all to work together more effectively (i.e. enhance productive efficiency).

2. The incentives internal to the enterprise should be such as to attract labour in order to maintain full employment and permit the expansion of the economy without undue inflationary pressures.

Recently, Martin Weitzman (1984) has actively promoted the idea of a share economy where labour and capital negotiate a share in net revenues rather than either a fixed wage rate (capitalist enterprises) or a fixed return to capital (debt-type co-operatives). Share arrangements, superficially, have the appearance of labour–capital partnerships; so would they prove attractive to the equitarian market socialist? I think not, but it is worth noting why.

The appealing feature of a share arrangement is that, since there is no fixed wage rate, an incentive always exists for capital (or managers acting as agents for capital) to take on new employees as long as the revenues generated for the enterprise are positive. By contrast, of course, the capitalist enterprise will only take on employees if the net revenues are greater than the negotiated wage rate, and the co-operative only if they are going to boost the average income of the existing members. This means that the share enterprise is by comparison labour-hungry. Even with a fixed component in the negotiated remuneration of labour, as long as this falls below what the negotiated wage rate would be, then share enterprises will still soak up additional employees.

Furthermore, if the revenues of the share enterprise drop (shall we say because of a fall in demand and thus in the price of the product) then, unless the marginal net revenue of an employee becomes negative, the share enterprise will still wish to maintain its level of employment, albeit at a reduced rate of remuneration. If there were other booming sectors in the market economy, then employees would be attracted to these; and this is of course desirable from the point of view of allocative efficiency. But, even in a general recession, share enterprises will maintain high levels of employment at reduced levels of remuneration. The share enterprise, in effect, shifts the risk of unemployment for labour to a risk of a reduction in its income.

However, established employees within a share enterprise will always eventually face a situation where, if the enterprise takes on new employees, the average income will fall (i.e. when the net contribution of a new employee is less than the average income of the existing employees). Then it is not in the interest of the established employees to welcome new ones, unless a reduced share contract is negotiated for them, introducing differentials in remuneration. This argument entirely parallels the one in connection with labour co-operatives found in Chapter 7. Thus, although there is an incentive for capital to take on new employees, who, in turn, wish to join the enterprise, the established employees have no such incentive. Indeed, unless the employment decision is in the hands of capital, then the desirable employment characteristics of the share enterprise are unlikely to materialize—certainly in the face of well-established norms about equal pay for equal work.

Clearly, Weitzman's much-applauded 'employment effect' will, as he fully recognizes, only become apparent in an enterprise securely in the hands of capital—something which is not consistent with the ethos of market socialism nor, in practice, likely to secure the enhanced productivity characteristics attributable to a community of interests.

Similiar reasoning applies to the capital-investment decision also. In this case, with a fixed-share ratio in the net revenues, labour will always be keen to take on new capital as long as its productivity is positive. Again, with eventual declining productivity of capital, unless capital can strike a flexible share-ratio contract with labour, there will be a disincentive for capital to invest in the share enterprise. Arrangements which allow flexibility so that labour and capital share the costs of new investment in the same ratio as their yields are a possibility; but this once again introduces conflict of interests into the very heart of the enterprise.

Our conclusions must inevitably be that share enterprises are not what we are searching for. In general, they do not generate a community of interests between capital and labour. To put it succinctly: labour wants more capital and less labour, capital wants more labour and less capital, and there is a conflict of interests between the established employees and those wishing to join the enterprise.

Meade (1986), in recognition of these problems, has developed the idea of discriminating labour–capital partnerships. This is not the place to explore his conception in detail, but such partnerships do appear to achieve the characteristics mentioned above.

A labour–capital partnership is essentially a form of enterprise which spreads the risk of variations in income between both the providers of capital and the providers of labour. Both possess share certificates which entitle the holder to a dividend in net income. Capital shares are tradable on the market though labour shares are not. The ultimate control of the enterprise is in the hands of the board of directors elected in equal numbers by labour and capital. New shareholders—capital or labour—enter the enterprise with an entitlement to a share in net income which appropriately discriminates in favour of the existing shareholders. This arrangement effectively overcomes many of the problems encountered in the Weitzman-type share enterprise. The existing capital and labour shareholders have an incentive to invite new shareholders (of either type) as long as their net contribution is positive. The labour–capital partnership is both capital- and labour-hungry.

Many variations on this basic theme are possible. For instance, both capital and labour may seek to strike a contract which only makes a proportion of their income variable and open to risk. There is thus within the framework of labour–capital partnerships a range of opportunities to share risks in a variety of ways whilst building convergent incentives. It is for this reason that the equitarian market socialist should be exploring the range of possibilities.

In a society where the positive freedoms in the relations of production are step by step made more equal, income differentials will decline. The introduction of various forms of labour–capital partnerships is also likely to reduce the spread of intra-organizational income (e.g. the Mondragon co-operatives). Greater 'equality' in consumption is thus already guaranteed. In so far as special needs and public goods have to be financed, then income tax (from labour and capital) will prove necessary. Income tax may not have to be very progressive; indeed, in the context of much greater equality, a non-distorting lump-sum tax may be seen as appropriate. The consumer will be protected, however, by the stimulation of competitive markets, but if certain technical monopolies are brought into a public sector then consumer representation on their boards will prove appropriate.

So, in an equitarian market socialist society, 'common ownership' comes to mean equally distributed initial endowments of equity carrying weakened property rights commensurate with increased rights going to labour, and 'popular administration' becomes equated with labour–capital partnership. Clause 4 takes on a new appearance, an appearance which in my view is far more likely to commend itself to an increasingly educated electorate than some tired old recipe about nationalization.

5

Planning in a Market Socialist Economy

Saul Estrin and David Winter

THE mechanisms of exchange in an economy are the means of transmitting information about current and future needs. This function is at least as important as the more familiar task of transferring goods and services from producers to other producers or to final consumers. A third element concerns the incentives which encourage economic agents to respond to the information they receive.

An 'efficient' exchange mechanism should perform these functions *without waste*. As we shall see, the concept of efficiency in this context has many aspects. Efficiency is not the only criterion, however. It is also important that an exchange mechanism generates outcomes which are consistent with the aims and objectives of society, or at least closer to consistency than any alternative mechanism.

There are numerous ways of organizing exchange, and no actual economic system would rely on only one. We distinguish some of these different mechanisms, emphasizing the importance of market as opposed to non-market forms of exchange. In this framework, planning becomes one possible form of non-market exchange. The key questions over the choice of exchange mechanisms concern the ways in which these different forms are mixed together, their relative importance, and the relationships that exist between them.

We would like to thank Michael Ellman, Louis Putterman, and Peter Wiles for helpful comments. Needless to say, the views and remaining errors expressed are our own.

Readers who wish to explore these ideas further might be interested in Cave and Hare (1983), Estrin and Holmes (1983), Hare (1985), Hodgson (1984), Nove (1972), Nove (1983), Nuti (1986), and Wiles (1964).

In every economy, one kind of exchange mechanism comes to be regarded as the principal means of allocating resources. This process will depend on historical circumstances just as much as on economic considerations. Other kinds of exchange mechanism will operate in areas where the principal method breaks down or performs badly.

It is conventional to associate markets with capitalism and planning with socialism. We believe that this is fundamentally misleading. Both capitalism and socialism use market and non-market methods of exchange. Both large corporations and the state use planning techniques in capitalist economies, and markets provide a back-up system in centrally planned economies. We will argue that, in a complex industrial society that wishes to adopt the goal of socialism, markets should be the principal form of exchange mechanism. In conjunction with other institutions, they can provide both the information and the incentives for an economy to allocate its resources in desirable ways.

However, they will not always perform well and much of what we have to say concerns identifying those occasions when we think markets will work badly. It does not follow that planning will necessarily be superior. There are other ways of producing desirable outcomes which are preferable to planning. The construction of an indicative plan may also yield benefits by reducing the amount of uncertainty facing economic agents. The experience of planning in France will be relevant here. A variety of government interventions will be needed, including in certain circumstances 'strategic planning'. By this we mean systematic intervention by the state to achieve particular targets in one or a small number of sectors of the economy, without regard to the consequences for the allocation of resources as a whole. The Japanese offer the best-known example of such an exercise. They provide, however, a method of implementing economic policy rather than making the choice of a different principal allocation mechanism for a socialist economy.

After defining essential terms in the next section, we turn our attention to assessing the performance of markets in the light of the aspirations which we assume market socialists will wish to adopt. In the following section we consider planning under market socialism, identifying areas where markets might be expected to perform badly. We then turn our attention to the theory of economy-wide central planning, complemented by a discussion of the experience in Eastern Europe. Our conclusions are given in the closing section.

DEFINITIONS AND DISTINCTIONS

Before proceeding with the main argument, a number of key definitions and distinctions need to be drawn. The first distinction concerns the ownership of the capital used in the production process, and of the residual surplus or profit that ensues. If private individuals (capitalists) own the means of production, this has undesirable consequences for the distribution of income and wealth within society, and for the hierarchical relations of power within the firm. Socialists of all persuasions have typically objected to these ownership arrangements on the grounds of fairness and equity and have proposed various forms of public ownership as an alternative.

In this chapter we will not consider the ownership issue. Markets, and for that matter planning, can be used in conjunction with a number of different kinds of ownership institutions, both private and collective. The discussion which follows is concerned with the way that transactions are made, rather than the rights to the stream of profits which may result from them. Because traditional socialists associate markets with capitalism, they assume that market socialism must suffer from many of the evils that afflict capitalism. This argument confuses means and ends. Markets are an exchange mechanism. They are a means by which certain economic activities are carried out. They are not an end in themselves. If the operation of markets is not inconsistent with a society which combines freedom, efficiency, and fairness, there can be no objection to the use of markets.

The second clarification we would like to make concerns the kind of activity we wish to denote as 'planning'. Planning from our point of view does not concern the activities of agents or firms in the economy; it only applies at the level of the economy as a whole. It consists of an effort to provide a comprehensive and internally consistent account of the development of the economy, attempting to cover all demand and supplies in the economy and their relationship to each other.

A popular use of the term planning describes all purposeful economic activity. You or I plan our careers, our expenditures, our savings, and our next holiday, etc. Any organization, large or small, is liable to plan its actions. This is not what we mean by economic planning, and it is not what most socialist writers have meant by planning either. For an individual or an organization to work out their activities systematically is very different from allocating resources across the whole economy according to a comprehensive plan. More importantly, the decision about whether or not to plan for an individual or an organization is essentially a voluntary one, and not really the subject of public policy. Planning one's activities will be a useful task in certain conditions and not in others, and the people involved are usually the best judges in such circumstances. Economy-wide planning, on the other hand, is pre-eminently a matter of public concern.

It is clear that a huge number of transactions (possibly the majority) take place within institutions—the household, the firm, or government. We refer to these transactions as non-market provision—that is provision which does not go through the external mechanism of voluntary trade but is entirely determined within the organization. For example, a single woman hires a cleaner each week; this is market provision. She marries and does the cleaning herself instead; this is non-market provision.

There are other kinds of non-market provision that we should mention. Firms often sign long-term contracts with their suppliers. The state frequently provides services such as education or health at zero prices. Although quite different, these methods of conducting transactions are clearly non-market in character, in the sense that either prices or quantities (or both) are fixed for comparatively long periods of time.

The problem of establishing the appropriate borders between one economic institution and another is highly complex. We assume that households should be formed voluntarily. The size of firms is of broader concern, however. Capitalist firms frequently find it in their interests to integrate vertically by buying their suppliers or expanding into retail markets. They also sometimes integrate horizontally, either to reduce competition or to amalgamate and form conglomerates that operate in a number óf unrelated markets. Firms can secure their inputs by arranging long-term contracts with suppliers, thus reducing uncertainties that result from unexpected fluctuations in market prices. On the other hand, some deliberately pursue a policy of buying at current prices or at best using short-term contracts in supply in order to benefit from the flexibility that such an approach allows. If the technology of production entails economies of scale, large plants and firms may result.

Market structure, the degree of uncertainty, whether the transaction is long or short term, and the kind of technology involved all play a role in establishing both the appropriate size of firms and whether the relations between them are governed by market or non-market institutions. It would be very difficult to declare *a priori* that one kind of exchange institution was better suited in any particular circumstances to any other. The best solution to this problem is to adopt a decentralized procedure and allow firms to settle these arrangements between themselves. This will be efficient in at least one way. The participants will have access to the relevant information, at comparatively low cost. Such decentralized decision-making also gives more freedom to agents.

A major reservation of market socialists to this *laissez-faire* approach arises from the dangers of monopolistic abuse. Monopolists can use their market power to exploit both consumers and suppliers. They also tend to be inefficient. This problem is discussed in more detail below.

It is of interest to note in this context that one of the principal characteristics of modern capitalist economies is the steady increase of non-market provision, particularly on the supply side. The recent spate of mergers in both the United Kingdom and the United States indicates the attractions of both vertical and horizontal integration. These may be the consequence of the peculiarities of the tax and finance systems of those countries, however, rather than of well-informed judgements concerning the appropriate boundaries between market and non-market provision. The rise of non-market relations has also been encouraged by the state. For example, in Britain, the National Health Service and the public education system are both non-market institutions.

We stress these distinctions at the outset because we believe that socialists have tended to lump together non-market provision and economic planning. As a result, it is 'socialist' to support non-market provision even when the resulting relations are blatantly exploitative, and to support central planning even when it is hopelessly inefficient. Market socialists, however, assert that market relations are superior to non-market relations in the large majority of potential trades. We shall discuss below the circumstances when they are not. Market socialism therefore entails the rolling-back of non-market provision to allow competitive forces to increase efficiency and choice and, at the same time, to root out monopolistic abuse.

It is important to stress that our vision also does not preclude a number of activities that have been associated traditionally with socialism. Here we would include the collective provision of consumption goods that socialists have often recommended. If citizens wish to use the democratic processes that are available to them to provide for goods collectively, this does not imply that any particular sort of exchange mechanism should be adopted for transactions concerning those goods. A democratic process could determine that public transport, football matches, opera performances, or health care are provided at subsidized prices. This may have consequences for the tax structure which should not be ignored, but it is not ruled out by our preference for markets as the prevalent exchange mechanism.

ASSESSING THE PERFORMANCE OF MARKETS

We begin this section by explaining why there is more to the option of leaving everything to the market than socialists have traditionally thought. But we fundamentally disagree with the Libertarian Right, which refuses to see the systematic flaws in the market mechanism. These throw into sharp relief the crucial need for state intervention in a market socialist economy. Such interventions are discussed piecemeal in this section, but we envisage them being brought into the co-ordinating framework of an indicative plan, as discussed in the following section.

In the first place, the market can provide incentives for people to act in a way that is socially desirable, without further central direction. Markets have a fairly 'natural' incentive system, arising from the fact that relative scarcities are reflected in prices, which also determine profits and costs. Assuming competitiveness, and that prices are a true reflection of scarcity, pursuit of individual interest is therefore in harmony with pursuit of the social interest. Moreover, markets are efficient in the use and transmission of information. For each product, decision-makers usually receive all the relevant information encapsulated in a single signal, the price. This moves up when the good is scarce and down when it is in abundant supply, without individuals themselves having to exchange any specific signals about quantities. Thus suppose there are thousands of producers or potential producers and millions of potential consumers of a particular commodity. Using a price system, each agent receives only one piece of information: the price. Using a quantity-based planning system, the number of messages that the planners would need to transmit will be very large. If we assume that transmitting messages in this way is costly, it quickly becomes clear that central planning will be very expensive. Decentralized decision-making is much cheaper. Buying decisions are left to those with the best information about what they want and production choices to large numbers of different suppliers, without the need for the information to be gathered and processed through some central agency.

Another important characteristic of markets when they are working properly is their competitiveness. There is a link between market forces, prices, and the existence of a large number of traders. When there are few traders on one side of a market, they have significant power to exploit the numerous

parties on the other side. In the case of monopoly producers—
for example oil suppliers in the mid-1970s or the European
airlines—this is done by raising the price. Where price behaviour is
regulated, the problem re-emerges in terms of lower quality of
service. Obvious examples of this are British Telecom and
British Gas (in both private and public guises) and other public
utilities. In welfare services there is a tendency for queues and
problems of low quality to develop.

Socialists have allowed themselves to become identified in the
public mind with monopolistic supply of this sort. If there are a
large number of potential sellers, so that buyers are not forced to
deal with one particular firm, these problems cannot emerge.
People can keep their options open about with whom they will
ultimately trade. Market forces favour the survival of helpful
suppliers and punish obstreperous ones. The reaction of citizens
as consumers, East and West, to the liberating aspects of market
relations suggests that this argument is an important one.

However, there are a number of occasions when market
allocations break down and when state intervention is required.
First there are a few goods, for example railways or the water
supply, for which the market can only support one or a few
firms ('natural' monopolies). More common is the case when a
small number of firms choose to suppress competitive outcomes
by acting in concert—a cartel to raise prices, reduce quality, or
restrict new entry. The problem of monopoly has long been
realized as a major drawback to developed capitalism. But there
is also a lot that the government can do about this problem.
Natural monopolies must be regulated, perhaps via public
ownership. If the monopolies are not natural, the enterprise
should as far as possible be broken up into competing units.
This is consistent with the objective of reducing non-market
provision, and opening new sectors of the economy up to
competitive forces. Similarly, multinational corporations
should be subdivided into their constituent national parts, so
that enterprise and national domestic objectives do not conflict.
The overriding principle is that the government should re-
introduce market relations whenever non-market provision has
emerged in pursuance of monopoly power.

In additional, the government must be ready to stimulate the entry of new competitors in monopolistic sectors, in the limit being prepared to set up productive capacity of their own, so that the threat of competition prevents firms from abusing their market position. The possibility that market shares, and profits, will be contested by new entrants if monopoly power is abused is a crucial element in competition policy.

The second major category of market failure is goods with spillover effects—when the benefits to those trading in particular markets are more than offset by adverse consequences to actors elsewhere in the system. As a general rule, markets tend to allocate too few resources to goods in which the spillover effects from traders to other parties are positive, like health care or education, and too much to products with negative external effects. The most obvious example of this is the use of production processes which damage the environment.

A familiar example is pollution. A soap powder manufacturer may ignore the effects on local fishermen of polluting the river upon which its factory is sited. One aspect of the problem here is that potential market relations between the two parties are absent. In principle, the manufacturer could be made to pay the fishermen for the right to pollute, or the fishermen bribe the industrialist not to pollute, but generally there is no market available upon which they could trade. A traditional solution would be a system of subsidy or taxes. One might also see the emergence of non-market provision: the two industries could be integrated into a single conglomerate organization in which allocative choices which could take pollution into account could be made at head office. The resolution of these problems could also be facilitated by the establishment of a consultative forum, such as an indicative planning process.

Thus, even for market socialists, the market cannot be the way to allocate goods with large spillover effects. The state will have to play a large part too—with taxes and subsidies, by arranging a consultative framework via the indicative plan, by regulating particular markets to prevent damaging side-effects, and by the direct provision of certain goods and services which private individuals have inadequate incentive to produce—law and order, defence, and so forth.

There is, in addition, one particular market in which serious allocative problems tend to emerge, and in which market socialist governments will want to intervene: the capital market. Decisions about whether to invest in one activity or another, or to use one technique rather than other, are much harder than decisions about, say, how many apples to bring to market. Capital accumulation relies on complex judgements about likely demand and cost conditions for many years into the future. They are necessarily a balance of expertise, technical knowledge, and guesswork. There are three distinct problems in relying on markets to make decisions of these sort. In the first place, the market may fail to provide sufficient or correct information to the investor about the future. It will be remembered that prices are the main source of information in a market system. Because, for the most part, we cannot trade today for goods to be produced or consumed in the future (future markets for currency and a few commodities like wheat are an exception here), the price in the future of material inputs, labour, interest rates, and the goods produced are simply not known when investors have to make their decisions. For example, in building a new power station which will not actually produce electricity for another fifteen years, costings and predicted profitability must be notional because the decision-makers just do not know how much, say, coal will cost, electricity workers will be paid, and the electricity board will charge fifteen years hence.

Because people tend to be rather cautious by nature, and because the uncertainties in investment can be so great, there will be a systematic tendency to underinvestment in a market system. Insufficiency of investment will slow the pace of economic growth, and restrict the improvement in living standards. Many consider underinvestment to be at the heart of poor British economic performance. Moreover, decision-makers will not merely underinvest across the board, but will bias their accumulation towards projects where the uncertainties are fewest and the risks least. Playing safe is of course a characteristic of banking and commercial companies whose role it is to fund investment projects. Yet it is often the riskiest projects—the internal combustion engine, the telephone, the micro-chip—which propel the motor of economic development. The market socialist state must therefore counteract these tendencies by intervening to provide firms with information about the economic environment through an indicative plan, and to foster both the general rate of accumulation and investment in relatively risky projects.

The capital market is also a place where the 'anarchy of the market' may be particularly costly. The market mechanism signals profitable opportunities through higher prices. But that signal is available for all to see, and, if entry barriers are low and set-up costs relatively small, there may be a scramble to meet demand by new firms, using a variety of methods to produce an array of slightly different products. Some of these firms will prosper, but most will fail. That is not necessarily a problem if a range of consumer tastes is thereby satisfied and if the waste involved in the initial over-stimulation of production is relatively small. But if the resources devoted to unnecessary investment are large, the economy pays a significant price in terms of duplication of effort for a small reward in terms of variety of the product. This suggests that there could be significant gains from the state using an indicative plan to co-ordinate investment in particular sectors or products.

Finally, it is obvious that the repercussions of investment decisions can be large for people far removed from the decision-making process. For example, purely commercial considerations predominate in deciding whether to build a new television tube factory in Wales or the south-east, whether to employ workers or to rely mainly on robots, or whether simply to shift all production to South Korea, even though the lives of people in these areas will be seriously affected. Investment decisions today shape the structure of production tomorrow, determining what the economy will be able to produce, what skills will be needed, and where workers will be required to live. Market forces saw the transformation of the rural north of England into an industrial powerhouse, at enormous human cost. It seems likely that, in the absence of intervention, it will see its reversion to an industrially peripheral, low-population-density area, once again at enormous human cost. Investment decisions also include the choice to decumulate capital—to scrap factories, equipment, and capacity which are considered no longer to be commercially viable. It is choices of this sort, each taken on sensible grounds from the perspective of the decision-maker but ignoring the cumulative effects on communities, regions, and the industrial structure of the economy, that call for public intervention. We discuss appropriate mechanisms in the following section.

Another serious deficiency of the market system is that price signals may need to be exaggerated in order to stimulate the desired adjustments to the quantities produced. As a result, prices can significantly overshoot their new long-run level after, say, a demand increase, and only gradually come down again. Thus the market system can produce excessive price volatility in the short run, with all the associated problems of uncertainty and waste. Such volatility will be particularly serious if, as with raw-materials producers, the price of the commodity is closely associated with the income of those who supply it.

The deviation of prices from their long-run levels is an important issue because it undermines the costless signalling function, which we previously viewed as one of the principal advantages of the market mechanism. The problem would not be serious if the overshooting did not tend to persist. The length of time for which prices deviate from their long-run levels depends on the rate at which quantities can adjust to price signals. If the required output changes can be easily implemented by the application of additional labour and materials to existing equipment, the adjustment will be relatively fast. If, however, they involve fundamental changes in the productive structure— new methods, new equipment, new product lines—the gestation lags will be longer. Then 'incorrect' prices —too high or too low—will persist for long periods of time, and lead to misallocations of their own: excessive adjustments in quantities because of the excessive movements in prices.

This is another version of the 'anarchy of the market' argument, with, for example, a demand increase stimulating price overshooting and, after a long period of insufficient supply, an over-response by producers. The general point is that, while markets may be excellent for fine-tuning responses to changing demand and technology, they may not be good at stimulating large, non-marginal changes in the structure of the economy.

The effectiveness of prices as a signalling mechanism usually depends on people being willing to respond to financial incentives. In fact, of course, individuals will be motivated by a variety of other considerations—traditional, communitarian, moral, religious, etc.—that conflict with material incentives. As a result, the response to market signals may be slow and weak. Even if financial gain is an important motivating force, the structure of incentives may mean that economic agents have radically to change their lives in response to market signals, for comparatively little reward. Such immobilities may make a certain amount of sense at the individual level, especially if price overshooting means that the perceived income gains or losses are less than they first appear. This will introduce further lags into quantitative adjustment and increase the lengths of time that the price will have to exceed its long-run level to motivate sufficient changes in supply.

The price mechanism will therefore provide excellent signals about shortages in products where supply adjustments are rapid and cheap. However, when there are lags in the adjustment of supply, large price fluctuations and overshooting may persist for long periods of time. This highlights a further role for the state under market socialism: intervening to dampen price fluctuations and the associated effects on incomes while directly stimulating quantitative changes in sectors where supply adjustments to demand changes are relatively slow. Once again, resolving problems of this sort will be a central function of the indicative planning agency.

Another major concern for socialists is the distribution of income and wealth. It should be stated at the outset that, provided the inheritance of wealth is severely restricted, and we recognize that this will be a difficult condition to meet in practice, an exchange system that will make some people temporarily wealthy is not, to us, an excessive price to pay for the other advantages of decentralized allocation.

It is often argued that, if resources are allocated via the market-place, then the frivolous tastes of the rich will receive priority over the basic necessities of the poor—supply for profit rather than 'social production'. There would appear to be two interrelated problems here, monopoly and distribution. A market socialist government would eliminate or severely curtail the former problem. In the absence of serious spillover effects and monopoly power, as we have seen, production for profit *is* socially desirable. Socialists are probably more worried about the fact that the price mechanism encourages producers to make goods for which there is a demand. Since the rich have more to spend, producers will devote more resources to satisfying their demands.

On the other hand, the poor have no choice but to spend their limited incomes on basic necessities. If food is expensive today, the poor cannot decide to buy something else instead. Nor of course can the rich. But they have a far larger cushion of resources to fall back on, and in any case spend a smaller proportion of their income on food. The consequence of this is that richer people tend to do their buying in more competitive markets than the poor. Markets work better for the rich. It is not surprising that numerous studies in the United States and other predominantly market economies find that 'the poor pay more'.

Though market socialists will clearly have a strong redistributive policy, concerns about the distribution of income are not that easily resolved in a market economy. Markets aggravate inequalities. With outcomes being uncertain, some people do very well and others very badly from trading. Moreover, as with gambling for high stakes, the people who do best tend to be those who enter the game with more resources at their disposal. The resulting inequalities persist from generation to generation. In most market economies today, the majority of the rich are rich because they started from a privileged position.

It is a basic tenet of market socialism that a redistributive system which largely abolishes previously inherited economic privileges will be put in place. Even so, in a market system a new group of relatively rich people will almost certainly re-emerge, either as a result of their own efforts or because of luck. Because wealth accumulates at a compounding rate, small differences in wealth are enormously magnified over time, particularly if the time-scale involved is so long that it spans generations. This highlights the crucial importance of breaking the inequality cycle by drastically hindering the wealthy's capacity to pass on their ever accumulating fortunes through the generations.

Some argue that the re-emergence of the wealthy will always undermine and eventually destroy a socialist society. Wealth, it is said, confers economic and political power which the rich will use for their own purpose, in particular preserving their own position at the expense of the poor. On this issue it is not clear that planning can do any better. Inequalities emerge because the economic environment is necessarily uncertain, so that, by luck or foresight, people can manage to be in the right place at the right time. With planning, the number of people with particular products at a particular time is restricted, increasing the potential for accumulation and aggrandizement from favourable starting-location positions. In fact, the diffusion of power and authority in a market system may act as a natural counter to, rather than reinforcer of, economic inequalities. Socialists are in part confusing the consequences of unequal distribution—that the wealthy do well in a market system—for its cause.

The wealthy élite will not be eliminated by a planning system; the form is changed to an élite of bureaucrats. This has happened in the centrally planned economies of Eastern Europe, where such groups successfully devote considerable resources to themselves. They are also in a position to pass their privileges on to successive generations. However, an important difference with rich capitalists remains. The planning élite's access to resources is based upon positions in the bureaucracy rather than on the ownership of wealth. Some might argue that it will be easier to sack bureaucrats than deprive wealthy traders of their assets, and thus deprive them of their powerful positions. Such evidence as exists, however, from the Philippines to Iran, from

Poland to Czechoslovakia, is far from conclusive on this issue.

The existence of short-term wealth inequalities under market socialism means that the problem of 'social production' emerges once again. Even if we can prevent inequalities from persisting across generations, some people will be (relatively) rich and others (relatively) poor at one moment in time. To prevent the system from penalizing the poor, a structure of welfare services will be required. It is very hard to envisage any kind of socialist society without a welfare state. Market socialism is no different in this respect. The kind of welfare state that market socialists would support is discussed in Chapter 8.

In summary, strong arguments related to decentralization and efficiency lead us to favour markets as the primary mechanism of resource allocation. However, competitive markets have serious problems from the socialist point of view, for example with regard to spillover effects, in the allocation of capital for large or non-marginal changes in the production structure, and in the distribution of income and wealth. Each of these problems delineates a role for government intervention. It is the willingness and the ability of the authorities to act decisively in these areas which help to distinguish a market socialist government from its capitalist counterpart.

PLANNING UNDER MARKET SOCIALISM

Market socialists cannot rely on markets to produce acceptable results on their own. A planning agency would be needed, though this would be very different from the planning agencies of the Soviet Union and Eastern Europe. Its job should be to patch up some of the informational and co-ordinating failures which we have already discussed. However, while we expect indicative planning to improve upon outcomes which would be obtained from a pure *laissez-faire* market system, it cannot be expected to eliminate all the problems we have mentioned. Moreover, it is important to stress that indicative planning is a valuable complement to, but not in any sense a substitute for, the market as the principal mechanism for allocating resources.

Indicative planning is a decentralized, and preferably demo-cratic, process of consultation and discussion concerned exclusively with plan construction and elaboration. The process provides a forum in which information can be pooled and in which diverse interest groups can confront one another concerning spillover effects. In itself, the plan does not contain an implementation procedure. It is left to individual agents to strike separate deals with one another within the planning framework, each deal enforceable like any other voluntary contract. Such a procedure contains rather more teeth then might at first sight appear, because one of the major actors in a market socialist economy is the state.

The historical experience of France offers us a model of indicative planning in a market economy. Though the political and institutional arrangements are far from ideal from a market socialist perspective—France had governments of the Right for much of the relevant period—French planning offers important insights into the practice of planning in a market economy. French planning was introduced after the Second World War. The planning agency is small and helps to co-ordinate govern-ment economic policy and to pool information about the economy for people concerned with the accumulation of capital in the public and private sectors. In contrast to planners in the Soviet Union or in other Eastern Europe countries, French planners have never had either resources to allocate or power to enforce their decisions. They operate by discussion, persuasion, and the provision of information.

At the heart of French planning is a complex system of consultation. Many thousands of people are involved in an exercise that, particularly during the 1960s, was a major effort to produce a broad social consensus on the nation's priorities in the medium term. For example, after 1959, the new Gaullist regime was committed to mobilizing the population and using state power to modernize France socially as well as industrially. The resulting Fourth Plan (1961–5) represented a genuine attempt to develop a consensus on social as well as economic issues.

According to the Planning Commissioner at the time, Pierre Massé, French-style indicative plans should be self-implementing.

This is because, he argued, they revealed a consistent pattern of future demand and supply for the whole economy, and because the key decision-makers themselves were involved in plan construction. Hence everyone knew that, if each merely did what they had promised to do, outcomes which were desirable, both individually and socially, would result.

In practice, a considerable arsenal of policy interventions, from subsidies and tax concessions to penalty clauses and credit restrictions, were also applied to make sure that agreements hammered out in the Planning Commissions actually did stick. But these were, at least officially, a matter of bilateral relations between the state and particular firms, and not the direct concern of planners at all. The planners merely provided a forum, and acted as its secretariat. French planning remained essentially a mechanism of information pooling and dissemination, gaining credibility from the quality and breadth of the information provided. It was also hoped that the process would counter the tendencies to risk-aversion and underinvestment inherent in the market mechanism. There is much in this that market socialists might try to emulate, though there may also be problems.

The provision of information is a sensitive and sophisticated task in an advanced economy. Consider the case of market production of chrome and cars, the former assumed to be a critical input and major determinant of the price of the latter. We can suppose that chrome producers know their own investment and future pricing plans, whilst car producers know how many cars they can sell at different prices. It would clearly be useful, in making investment decisions, for car producers to know the future price of chrome, in order to evaluate likely future demand. In the absence of future markets or indicative plans, such information is not available.

Indicative planning creates a forum for the two parties to meet and exchange the relevant information, and in this way to reduce this uncertainty. However, chrome producers do not know everything that is relevant in determining the future price of chrome, for example, they may not incorporate the effects of discovering major new chrome reserves, or the development of a new cheaper chrome substitute, in their price forecasts. Thus pooling all the information that everyone in the economy knows does not necessarily eliminate economic uncertainty about what will happen. There remain some things, for example natural disasters, wars, famines, and the like, which are in principle unknowable.

Pooling information eliminates a certain sort of uncertainty, usually called 'market uncertainty', but leaves a residual of 'environmental uncertainty'. This limits the scope of indicative planning. Forecasting errors in the face of major unforeseen events—like the oil price increases of 1973 and 1979—have in practice dented the reputation of French planners. But a more important factor in explaining the recent demise of the process is the ideology of the anti-planning Right which has been in power from 1974 to 1981 and from 1986 until 1988.

It should be no surprise that indicative planning cannot eliminate all economic uncertainty. It is a fallacy of central planning to believe that outcomes can be achieved regardless of circumstances, and therefore to set plan targets without reference to conditions during the period of implementation. Environmental uncertainty highlights the form of planning appropriate for an advanced industrial economy. Fixed planning periods, typically of five years, should be replaced by 'rolling plans'— plans in which details for the early years are well specified but intentions for later years become sketchier and sketchier as the distance from the starting-period increases. Moreover, traditional fixed targets should be replaced by contingent ones. There would therefore be a number of plan variants, each associated with different assumptions about the likely configuration of the key variables to which the economy was particularly sensitive— the exchange rate, world interest rates, etc. Contingent planning is far removed from the traditional socialist notion of plan targets, typically set high to be achieved by Herculean effort. But, by highlighting the fragility of forecasts in a market system, they more accurately reflect the informational needs of a market socialist economy.

Indicative planning can also allow wider social involvement in the allocation of resources and internalise spillover effects. For example, one can envisage regional representatives intervening to prevent the adoption of a plan in which, under all likely contingencies, unemployment levels in the North of England grow. In addition to civil servants, industrialists, and unions, planning would have to involve consumer and regional representatives, outside experts, and, where necessary, the representatives of special interests often ignored in the decision-making process—such as the fishermen about to be polluted by soap powder manufacturers in our previous example. As Paul Hare has suggested (1985), this will involve considerable decentralization of the planning process, with planning councils linked at national level with the government and large corporations, and at local level with production establishments, local authorities and the community. It is to be hoped that planning of this sort would be democratic, in the sense that information about possibilities and objectives would flow up from the planning committees to be collated by the centre, rather than priorities being imposed from above.

Indicative planning therefore offers a decentralized and potentially democratic version of planning which can improve the functioning of markets, without threatening to displace them as the principal allocation mechanism. Its primary contribution is intended to be improving economic efficiency, rather than directing economic activity. It operates through the provision of information, and its effectiveness in large part depends on the sophistication and usefulness of the data made available. The contribution of indicative planning to economic welfare may appear small to traditional socialists, who compare it with the directive organs of central planning. Viewed from the perspective of a market economy, the contribution of indicative planning is much larger.

THE THEORY OF ECONOMY-WIDE CENTRAL PLANNING

We now turn to an alternative method of allocating resources: economy-wide central planning. The case for market socialism must rest on more than the advantages of decentralized allocation. From a socialist perspective we would argue that the benefits of markets probably outweigh the costs. Nevertheless, we do not deny that these costs can be large—particularly in terms of investment, when non-marginal structural changes are required, and in the pursuit of egalitarian distributions of income and wealth. Indicative planning can reduce but not eliminate these costs. The case for market socialism must also rest on the failures of centralized allocation. It is these that we consider in this and the following section.

Central planning typically has two phases: first, when all the activities of the various actors in the economy are brought into line with each other, in order to pursue some agreed end; second, when the desired allocation is implemented. The first stage—plan construction—entails a transmission of information about tastes, technologies, and opportunity costs between all agents. The second stage—plan implementation—may entail further co-ordination because the plan may have to be simplified to be implementable. For example, some parts of the economy may have to be grouped together or even omitted if they are peripheral. In contrast with indicative planning, the bulk of the effort in economies that are popularly regarded as planned arises in the implementation phase.

As we have already pointed out, because the future is uncertain, planning organizations often find that their plans go wrong. Planners may be ignorant of current economic conditions, and future circumstances are always uncertain. It is a recurrent feature of economic plans that they frequently fail to achieve their goals. To see why, it is necessary to examine the process of economy-wide plan construction more closely. For a plan to succeed in the stated time period, it has to be, at the very least, feasible. It must be possible, with the resources that the plan allocates to the planned tasks, to achieve the envisaged ends. Most plans are at least intended to be feasible. Most will have elements which, in the event, prove to be unfeasible, because of unforeseen circumstances. Some improvisation will always be required, and what happens during the implementation of plans that are not feasible is of considerable importance. It plays an important role in the success or failure of the planning process.

Central planners, like indicative planners, find it hard to construct feasible plans because the future is inherently uncertain.

To make the relevant predictions, planners need to be exceptionally well informed, not only about present economic conditions but also about the relationships that determine future conditions as well. This is a very difficult and costly task. One way of making it simpler is to arrange units of production into a small number of large firms. It is relatively easier to establish, for example, how much steel or how many computers the economy can produce if there are a small number of steel or computer producers. This tendency for planners to favour and encourage a small number of large firms has undesirable consequences. Just as large firms will tend to behave like monopolists in a market environment, in a planned system they will acquire and use to their own advantage political power. They will become a powerful vested-interest group that will advance their own interests at the expense of the rest of society. They will also act as a powerful obstacle to potential reformers.

Planners will thus need to devote considerable resources to acquiring information. Planning can only be as good as the planners themselves. From the point of view of feasibility, planning will be at its best in highly specific areas where planners will be well informed, where the spillovers and interactions with the rest of the economy are not considered to be important, and where the processes which govern future events are well understood. The less uncertainty about the future the better.

Once President Kennedy had decided that the United States was to place a man on the moon by the end of the 1960s, NASA was able to begin a very effective planning process to meet one specific goal. Its effectiveness, it should be noted, was partly the result of NASA's ability to acquire more resources from Congress if it needed them. There was in actuality almost no resource constraint, which meant that the question of feasibility was much easier to resolve.

Ends need to be clearly defined. One of the advantages of Kennedy's request to NASA was that it was quite unambiguous. Formulating specific, unamibiguous ends of this sort for the whole economy can be very difficult. The planners may be able to decide how much steel should be produced. But questions then arise concerning the quality, thickness, and strength of steel produced. Planners may wish to leave such details to the steel producers. But it is not clear that they will have the incentive or the information to produce steel of the required quality, thickness, and strength. Markets resolve these difficulties easily. If you are planning the whole economy, it is a serious problem.

Feasibility is only one property that a good economic plan should have. Perhaps an even more fundamental requirement is that the goals of the plan should be, in some sense, desirable ends for the society as a whole. How should planners decide on what the goals of their plans should be? In a national economy, the objectives of a plan can be hard to determine. There is at the moment on the Left in Britain a feeling that the country should invest more in manufacturing. The question then arises as to which kind of manufacturing industry? Should we have a shipbuilding industry? If so, how large should it be, etc.? These are not easy questions, and planners are not necessarily the best people to give the right answers.

It helps the process of implementation, as well as being desirable in its own right, if the ends command a consensus. Planners, in the past, have frequently ignored the preferences of the bulk of the population. Central planning is associated with authoritarian, if not tyrannical, political systems, in which the preferences of a small élite of planners and their political masters are imposed on the rest of the population. Is it necessary for planning to have an authoritarian element? Is it possible to envisage planning procedures that would be acceptable in a democratic society?

The answer to this is that, while, as we have seen, economy-wide plan construction can be part of a democratic process, it is hard to envisage effective plan implementation that allowed or encouraged a wide diversity of views among the participants. Central plan implementation naturally takes place within an essentially authoritarian hierarchy. Unless the tasks are so simple that one or two people can perform them all, teams will have to be formed and their activities co-ordinated. It is difficult to see how such teams can operate on a genuinely democratic basis to implement predetermined targets.

Capitalist societies often adopt planning during wars. Here the aims of the society can be clearly stated. The means to achieve them are also usually fairly clear. The aims command widespread agreement, and there is little resistance to authoritarian methods of implementation. Resource constraints can be temporarily relaxed. Wars even provide an incentive scheme (military discipline and the fear of losing) in contrast to the weaknesses of planned economies in this regard.

It is possible to envisage pluralist procedures being incorporated into the plan-construction stage in peacetime. While the ends of a plan are being discussed, a number of different views and interests can be incorporated into the planning process using democratic, judicial, or consultative procedures. In the final plan, however, a clearly defined set of ends will have to be produced. These will have been formulated by specialist planners, even if they take into account the views of other people. Some non-specialists will inevitably find that their views have not been incorporated into the final plan. They may feel aggrieved but, if they are involved in implementation, they will have no alternative but to co-operate.

It is worth comparing planning and markets on this point. Markets give an advantage to the rich, in the sense that the preferences of the better off can usually be expected to have a larger effect on prices simply because they have more money to spend. But, apart from differences in income which can be reduced by taxation, all consumers will influence prices equally by the pattern of purchases that they make. A democratic planning procedure would give every voter an equal voice in the determination of the plan objectives. This may be considered more desirable than allowing those with high incomes to have greater weight. On the other hand, those with strong interests in certain plans tend to use their influence in order to secure the formation of plans that they consider to be desirable. Interest groups are a persistent feature of all societies in which governments have a large amount of economic power. Planning institutions provide them with easy opportunities to apply pressure. Socialists have to recognize that even democratic planning procedures would be vulnerable to the blandishments of interest groups.

As we have already stressed, it is in the nature of plans that they will frequently be unfeasible. When this occurs, two rather different kinds of things can happen. First, the ends of the plan may simply be revised to incorporate the new information regarding feasibility. In all centrally planned economies, plans are frequently revised in this way, so that the final plan corresponds with what has actually happened, and is 'successfully' implemented. Plan revision of this kind usually requires reference back to those who originally constructed the plan. If there is a centralized hierarchy involved, this may be difficult to do.

Second, those involved in plan implementation can attempt to improvise in order to come as close as possible to plan fulfilment. How and to what extent they will do this depends on the nature of the plan itself, the kind of economic environment in which they find themselves, and the structure of incentives that they face. In most centrally planned economies, special kinds of dealers emerge whose task is to attempt to make corrections for mistakes in the original plan. They acquire information on the availability of spare parts and surplus inputs. They bring together firms who can engage in beneficial exchanges of inputs or outputs which will help each to fulfil their own plans. These kinds of dealers play an indispensable role in any kind of planned economy.

From our point of view, their interest lies in the fact that they are essentially *market* operators. When the economy-wide central plan breaks down, firms are forced to trade on markets. If the economy has not developed and supported its market sector, then this kind of improvised trading will be more difficult to carry out, and the mistakes of the planners will have, as a result, more serious consequences.

The kind of planning that we have been considering is a social process in which, at the plan-construction stage, planners elicit information about feasible possibilities and different preferences from all other agents in the economy. Those who work in the relevant areas being planned attempt to make the planned ends a reality during implementation. So far, we have not discussed whether it is in the interests of these various participants to carry out efficiently the various activities which a planning process will require of them.

Sometimes socialists assume that people in a socialist society will naturally carry out their work without any incentive for doing so. If workers are not alienated from the process and institutions around them, they should be willing to work without a system of direct financial rewards. It is true that, under certain circumstances, loyalty and solidarity can be powerful and effective in securing harmonious behaviour. However, it is most unlikely that, outside extreme crisis conditions, an economy can perform effectively in the long run without a pattern of direct financial incentives. Planning has obvious deficiencies in this regard. Whereas, under a market system, profits usually (but not always) provide a socially desirable measure of efficiency, and can thus naturally form the basis of an incentive scheme, no corresponding measure of success exists in a planning system.

The obvious indicator of success in a planning system is plan fulfilment. In practice it presents numerous difficulties. A plan must be specified in an exceptionally detailed way if plan fulfilment can be achieved only by producing efficiently. Usually, it is possible to fulfil the plan, but at the same time hire more labour, run down stocks, or engage in other activities that circumstances allow but that have not been fully anticipated in the plan. It may be possible to exceed plan targets, but an incentive scheme based on plan fulfilment will not encourage this. A plan may be viewed as altogether impossible by those whose task it is to implement it. Again plan fulfilment provides no incentives in this case.

The problem becomes worse when the process of plan construction is taken into account. During this phase, planners will wish to have access to information about feasible possibilities as well as the preferences of other members of the society. It is highly likely that producers will have a strong incentive to distort any information that they give planners in order to secure plans that are easy to fulfil. Similarly it may not be in the interest of various agents to reveal their preferences for certain ends, if they feel that, by not doing so, they will advance their own private interests.

It may be useful to end this section with a brief summary of the points we have raised so far. We began by distinguishing between plan construction and plan implementation. In the plan-construction phase we argued that feasibility was an important property of an economic plan. A number of points follow from this.

1. The construction of feasible plans requires that the planners have access to costly information.
2. Planners find it easier to construct feasible plans if the degree of uncertainty is relatively small.
3. It is easier to construct plans in highly specific parts of the economy where the interactions and spillovers with the rest of the economy can be ignored.
4. Planners tend to favour large firms as they simplify the planning task.

The fourth point applies to both construction and implementation of plans. In addition we argued that:

5. Plans are easier to implement if there is consensus about ends and if those ends are clear and unambiguous.
6. Plans are easier to implement if there is a weak resource constraint.
7. Plans need markets to enable successful improvisation to take place during implementation.

Finally we considered the relationship between planning and incentives, arguing that:

8. Planning systems find it hard to incorporate successful incentive systems to encourage participants to reveal truthfully their preferences during the construction and to implement plans efficiently.

These points do not by themselves establish the comparative advantages or disadvantages of planning and markets. They make an interesting contrast with the behaviour of markets which we have already discussed, and suggest that there are certain historical conditions under which central planning can be expected to perform comparatively well. But these conditions are likely to be quite different from the ones facing complex industrial, democratic societies which rely to a considerable extent on world trade. We now turn to an examination of the experience of planning in the centrally planned economies of the Soviet Union and Eastern Europe, and of the Japanese experience of strategic planning.

THE EXPERIENCE OF CENTRAL PLANNING

The first socialist group to establish themselves in power were the Bolsheviks. They did not come to power democratically and, after winning a particularly brutal civil war, made no attempt to institute democracy. Instead, under Stalin, the Soviet Union developed into one of the century's most savage tyrannies, and, despite current reforms, its government remains a largely dictatorial political institution. In view of these facts, democratic socialists are understandably reluctant to look to the Soviet Union or its Eastern European satellites either as a model for socialism in Britain or as a justification for various socialist ideas.

We would agree that none of the political institutions and practices in the Soviet Union and similar states should form part of socialist Britain. However, we believe that the experience of the Soviet Union and its Eastern neighbours in planning can illustrate the strengths and weaknesses that we are discussing.

The Soviet system of central planning, as it developed under Stalin, is based around the formation and implementation of annual plans rather than the better-known strategic five-year plans. The annual plans are constructed by planners who are prominent members of a large hierarchical bureaucracy. Beneath the planners are industrial ministries and departments, who in turn oversee enterprises. Above the planners, there exists a political structure, in particular the Council of Ministers.

The current Soviet regime is attempting to reform the planning mechanism. It is not yet clear what form these reforms will take. However, at the time of writing, Soviet planners still concentrate on planning on a year-to-year basis. By doing so, they can fall back on the levels of output and the corresponding inputs already achieved. Much of their work can be described as planning from this 'achieved level'. This simplifies their task, but it also creates difficulties of its own.

Each annual plan is constructed during the year before it is implemented. The planners will therefore not know the levels of production of the current year. They will have to wait until the end of the year before production totals are known, and probably even then there will be a lag before the relevant data can reach them. By this time, of course, the next plan is being implemented, so planners have to make estimates of current performance as well as of future possibilities. In order to do this, they naturally engage in a consultation process with the ministries beneath them, and they, in turn, with the enterprises beneath them.

This process of consultation involves a large element of political bargaining. Enterprises on the whole wish to persuade the planners to provide them with a plan that they will find relatively easy to implement. In this way they will guarantee the bonus paid for plan fulfilment. The industrial ministries will also want their own enterprises to fulfil the plan to prove their bureaucratic competence, so that they will abet the enterprises in their bargaining strategy. The planners' response to these tactics is to attempt to set 'taut' plans: plans that are achievable but only at the limits of the enterprises' capabilities. In this way, the planners hope that some kind of productive efficiency will prevail.

Judging the correct degree of tautness is a major part of the planners' art. If a plan is too taut, the enterprise will feel that it is not possible to fulfil it, and will not attempt to do so. It may even decide to produce at high levels of output, and disguise this fact by shifting some of this year's output into next year's production. This can then be used to help fulfil next year's plan. If, on the other, hand a plan is not taut enough, the enterprise has no incentive to produce efficiently at all.

Many of the most characteristic features of centrally planned economies are the consequence of tautness. First, they provide one of the chief advantages of central planning. If for any reason there are unused resources in the economy, taut plans force enterprises into wishing to employ as much of them as possible. This suction effect does not mean that unemployed resources are used as efficiently as possible. But any use of a resource that has previously been idle is clearly an economic gain. Much of the high growth rates that centrally planned economies attained in the 1930s and 1950s came from this source, the unusued resources being frequently found within the agricultural sector. Labour shortages rather than unemployment are generally the problem for a centrally planned economy.

Taut plans absorb surpluses and create shortages. They encourage what has been described as extensive rather than intensive growth. In other words, centrally planned economies have tended to grow as a consequence of more resources being deployed in production, but not as a result of increases in efficiency. They are also characterized by shortages.

A further factor is at work here: the preferences that planners have adopted in constructing their plans. Since no democratic centrally planned economy has ever existed, central planners have adopted the preferences of their political masters, often at the expense of the wishes of the population. In particular, all centrally planned economies have given a high priority to economic growth, especially in the manufacturing sector. This has entailed high investment rates at the expense of current consumption.

Central planners' preferences for growth, which have sometimes been taken to almost absurd extremes, are often portrayed as an unconsidered prejudice in the West. However, there is an historical justification for this policy. Central planning was in large part adopted in the Soviet Union so that the economy could catch up with the Western capitalist powers. In the late 1920s the Soviet economy was growing more slowly than a number of Western economies, and Stalin wanted not only to increase the growth rate but to increase it sufficiently to catch up with the West. This desire to compete economically was partly motivated, in the 1930s, by the (as it turned out justified) fear of invasion and, in more recent decades, by the Cold War. The

Soviet Union has had some success in catching up. In the late 1920s average incomes were probably less than a quarter of the American level. Now they are about one half, despite the rather different experiences of the two countries during the Second World War. The gap has been widening again in recent years.

The politicians and planners may be able to justify their preferences for growth and investment in this way. Nevertheless these preferences often conflict with those of the population as a whole. Consumers are often painfully aware that their own needs receive a low priority in centrally planned economies. Rarely is there much investment in stocks or in retail services, so that, even when supply conditions are favourable, consumers find that to spend their income they are required to wait for long periods for durable goods and queue in shops for food and other consumption goods. This leads to a cycle of low morale, a lack of enthusiasm at work, and, as a result, poor standards of service and quality of workmanship.

During the period of implementation, when, as we have seen, a number of enterprises will inevitably discover that they are unable to fulfil their planned targets, a further process of bargaining to revise plan targets takes place. Political influence and group pressure will be important here. Enterprises will also use their network of dealers to attempt to obtain, by unofficial means, scarce inputs which will enable them to fulfil the plan.

This wheeling and dealing often involves illegal or semi-legal activities which the authorities reluctantly permit. They are aware that, without it, few plans would be fulfilled. There are numerous stories of the lengths enterprises will sometimes go to find and transport scarce inputs many thousands of miles in order to fulfil their plan. Obviously there are cases when one small spare part is worth thousands, if not millions, of roubles to an enterprise if, by obtaining it, it can fulfil its plan, and thus earn fulfilment bonuses. The fact that some enterprises will go to great lengths and pay very high prices in these black (or grey) markets encourages speculation and further undermines a proper price system. These activities also aggravate the perpetual state of shortage. A centrally planned economy acquires features which are arbitrary, even irrational. Socialists attack the 'anarchy' of markets, but there is also an 'anarchy' of central planning.

Although shortages, waste, and poor quality outside priority sectors have become the norm, central planning in the Soviet Union, particularly of the 1930s and 1950s, has some achievements to its credit. The planned economy performs in a desirable way when there are unemployed resources available. It can also, as we would expect, respond well to occasions when high priority is attached to clearly defined goals, which can be achieved without resource constraints. On the other hand, central planning has crucial failings as well. Poor resource allocation, low efficiency, bad quality of work, and the burden of a large authoritarian bureaucracy.

It is not surprising that, shortly after the death of Stalin in 1953, attempts began to be made to reform the essential structure of the centrally planned economy. In the last thirty years numerous reform proposals have been made, and continue to be made to this day. There have been a number of attempts to implement them, nearly all of which have failed. The chief candidate for change is, understandably, the centralized bureaucracy itself. If only decision-making could be taken at a more decentralized level, a lot of the obvious faults of central planning might disappear. There are a number of reasons why this appealing argument turns out to be wrong, and they well illustrate the problems of planning.

The first difficulty is that the central bureaucracy may not wish to be reformed. Reform in the Soviet Union appears to have been persistently frustrated in this way. Khrushchev's fall from power can be attributed to this factor. Gorbachev's direct appeals to the Soviet people can be seen as an attempt to circumvent the bureaucracy. It is too early to say whether he will succeed. But in the past the bureaucrats have ensured that reform is so half-heartedly carried out that it leaves almost no impression on the economy of the Soviet Union.

Alternatively, if substantive reform is carried out, this too can lead to economic disaster, as in the case of Poland. At the moment, only in Hungary does there appear to be a chance, however slim, that reform may eventually lead to a more efficient decentralized economy. The Polish example clearly illustrates a number of the problems. We know that, if economic decision-making is to be delegated to enterprises, then they must have an incentive to operate efficiently. So far the only decentralized system that provides a possibility of this is competitive markets with market clearing prices. In Poland, in the early 1970s, a substantial number (but not a majority) of Polish enterprises were given, not complete, but a fair amount of freedom. Some years later, in 1982, there was the biggest fall in production of any industrial economy since the Second World War.

Many factors were at work here, but the enterprises which had been reformed played an important part. Because the prices of goods that were not sold on consumer markets—that is, of goods that were inputs into production—were used largely for accounting purposes, they rarely provided accurate information about the relative scarcity of inputs. Firms responding to these prices were therefore making the wrong decisions. The enterprises themselves had been directed to maximize both profits and value added—aims which often conflicted, especially if, as in this case, the enterprises had considerable monopoly power. This was itself the result of the planners' encouragement of mergers, in the mistaken belief that a small number of large enterprises were easier to control.

In addition, Poland in the 1970s decided to increase its trade with the West. This was at a time when Western banks were sated with funds from the OPEC states. They were willing to make loans to almost any borrower. It is not surprising that the reformed Polish enterprises increased wages rapidly, imported imputs lavishly, and began a series of vast investment projects, funded by Western banks, many of which were never completed. The result was a huge balance of payments deficit funded by the West, inflation, and ultimately a disastrous recession. This the government blamed on the trade union, Solidarity, but the strikes of 1980 and 1981 were far too short to have had such drastic effects. The causes of the crises lay in the policies of the previous decade.

Poland provides a particularly graphic example of the difficulties of reforming a centrally planned economy. With hindsight, it is possible to see that the Polish government made almost every possible mistake. The Hungarians, on the other hand, have adopted both a more comprehensive and a more cautious policy. But they too have made mistakes. As so frequently with planners, before the main reforms were introduced in 1968, they had merged enterprises into a small number of large units. When these were given the freedom to make their own decisions, they too had considerable monopoly power.

The Hungarian government has had two kinds of response to this problem. One was to reimpose control and prevent monopolistic enterprises from raising prices to increase their profits. Since they had abolished the annual production planning apparatus, it was difficult to go back to exactly the same pattern of central control as before. The government elected instead to manipulate taxes and other 'parameters' that governed the finances of the enterprises as well as to control prices directly. This kind of control is closer to the manipulation of markets that we believe would be preferable under market socialism.

The second method adopted by the Hungarian government was to use the world market. Hungarian prices were to be the same as world prices. Hungarian enterprises were encouraged to trade on world markets, where they would be sufficiently small to act as competitors rather than as monopolists. The Hungarian government is now also encouraging small-scale private business and co-operatives financed through (relatively) independent banks. Large-scale investment still remains controlled at the centre, but moves to institute decentralized market mechanisms in the investment sector are now under discussion.

Hungarian reform has not been without numerous setbacks. The temptation to reimpose central control when things go wrong is still very strong. The use of world markets as a method of imposing efficiency has led to the Hungarian economy becoming vulnerable to the fluctuations of world prices. Little progress has been made in forcing loss-making enterprises into bankruptcy. The overall growth of the economy has recently been worse than that of some of its centrally planned neighbours, and the weight of international debt threatens to push the economy into crisis.

Both the Hungarian and Polish experiences show that, once a decentralized market economy has been replaced by central planning, it is very difficult to recreate it. The Hungarians have, in a sense, been forced to import the market mechanism. The costs and dangers of recreating such a mechanism after it has been abandoned can be enormous, as the Polish experience illustrates.

In the end central planning has few advantages over market systems. Moreover, by essentially destroying the market mechanism, it deprives the economy of a vital asset. Quite apart from the political costs, this is too high a price to pay for achieving full employment and the high rates of growth derived from the use of unused resources.

There is one other historical example which is relevant to our argument. Japan is a notably successful capitalist economy that has also used a kind of planning since the Second World War. Naturally, this has attractions to those who wish to emulate Japanese achievements. We find that the Japanese method of planning, like central planning, has crucial drawbacks, even though it exploits one of the areas which we identified as being particularly advantageous to planning—namely that planning can succeed in highly specific sectors of the economy. The Japanese experience paradoxically shows why, in the end, this is not a good foundation on which to base a planning procedure.

The principal agency involved in Japanese planning is the Ministry of International Trade and Industry (MITI), which has sought to pick the sectors in which Japanese investment should concentrate, thereby closely guiding the pattern of industrial development and growth. MITI therefore chooses to promote a particular sector of production, for example shipbuilding after the Second World War, or more recently electrical goods, electronics, and computers, and then co-ordinate an array of economic policies to ensure its rapid development. Import controls can be employed to protect domestic producers against foreign competition, particularly in the early years of product development. MITI offers funding on favourable terms for investment, technical licences, or patents, and sponsors some research directly. It therefore offers target sectors an integrated package of temporary protection and subsidy for investment and research, while stimulating domestic competition in these products by spreading its favours widely. Groups of firms are able to build a secure domestic base from which to launch co-ordinated assaults on the world markets for specific products— cars, stereos, televisions, computers. The success of the Japanese economy has encouraged many on the British Left to favour strategic planning as the model for planning in the United Kingdom.

While one cannot rule out the use of strategic planning in all circumstances, particularly as an effective tool to mobilize resources for a specific sector, it sits uneasily with our vision of market socialism. In the first place, there are a number of special circumstances about the Japanese economy in the immediate post-war period that do not apply to a developed Western economy such as Britain, some forty years later. The Japanese problem after the war was to choose a particular structure of production and to import state-of-the-art technology in order to build up a domestic production base. The choice of sectors was relatively straightforward; there was little domestic opposition to promoting one branch rather than another, and other countries were willing to accept Japanese import controls. The political system and its associated bureaucracy were too weak to interfere with the planners.

But it seems inconceivable in any medium-scale contemporary market socialist economy that domestic protection on this scale could be implemented without retaliation. Moreover, the choice of sectors is now far less simple, and the possibilities for effectively promoting sectors have been reduced in the more sophisticated and heterogeneous economies of today. Everyone would support a policy of picking winners, but there is no reason to suppose that the government is better able to do this than any other private entrepreneur. Public bureaucrats are further away from market realities, and are not offered the same material gains from success. And since there will always be uncertainty, the planners might easily pick losers instead. It is tempting to argue, in observing previous British failures to develop new products, that the problem lies in an unwillingness to invest in economic certainties. Such a view takes advantage of hindsight; few would have predicted twenty years ago the scale of demand for home computers, video recorders, or even air travel. It seems more likely that firms failed to invest because the options open to them were so varied and complex that they decided to wait and see what happened elsewhere, and then missed the boat.

This is a serious problem, but not one that necessarily will be resolved by the government targeting particular sectors on the basis of its own judgement and spending resources which may prove to be wasted. It is better to make such decisions on the basis of the maximum information available, within the framework of a coherent indicative plan. This would allow the national economy to spread its risks against various contingencies in consumer demand, research and development, and world trade.

The third major problem with strategic planning is its essentially undemocratic nature. The underlying concept is the choice by a group of 'experts' of the best conceivable path for economic development, to be implemented systematically by rewarding favoured sectors and punishing the unfavoured. This of course makes more sense in a hierarchically ordered society. It depends crucially on the weakness of central government, with its orientation towards the short term, as against technocrats entrenched within the civil service. For example, in the British context it would require that the short-run concerns of the Treasury and the Cabinet be subordinated for long periods of time, say ten or twenty years, to the strategy of an élite planning ministry. This would require fundamental, and probably unacceptable, changes in the way that our political processes operate.

Also worrying for a democratic socialist is the innate tension between strategic planning and any form of decentralized or democratic planning. In a relatively less developed economy, it is easier to favour particular sectors because this is not obviously done at the expense of others. Once the economy is more advanced, prioritization of one group means holding back another, and in a democratic system those that receive a low priority can be expected to protest. For example, strategic planners might decide that the future of Britain rested with genetic engineering, which was best located in the south-east to be near the agricultural markets of Western Europe. Such favouritism would be opposed by others attempting to spot the industries of the future—in lasers or computers—even if there were commercial logic to the planners' choice. The only democratic solution is to involve these other parties in the determination of a national strategy, in which case the whole strategic planning exercise expands until, in the end, it becomes the kind of comprehensive indicative planning we would in fact support.

Thus there may be occasions when strategic interventions in particular sectors or regions can be justified—particularly when ends are agreed and simple, and resources are freely available. But this does not mean that contemporary socialists should point to the post-war Japanese experience as an alternative to markets in the allocation of resources. Japanese planning is highly specific and has many features inappropriate to a decentralized socialist system.

CONCLUSION

Any economic system will adopt a particular type of system that becomes the principal mechanism of exchanging goods and services. Other mechanisms may coexist, but they will be supplementary to the principal form. These supplementary mechanisms will operate in conditions and under circumstances when the principal form has ceased to function efficiently. Thus, in economies where central planning is the principal form, it is frequently impossible to implement plans as they are formulated. In this case either the plans themselves can be revised or a market-type mechanism comes into play. Since market relations are often either not encouraged or illegal, these black or grey market transactions take place under conditions which do not make markets particularly efficient. Nevertheless such market transactions are more efficient than the available alternatives in the context of a centralized planned economy.

The chief problems of central planning are not only unfeasibility, tautness, and the resulting incoherence and inefficiency. Central planning institutions also suppress and damage the market mechanism. If you want to observe the most inefficient and corrupt abuses of market power, you will find them in planned rather than in market economies. Of course other kinds of supplementary exchange mechanisms come into play in centrally planned economies. The preference by planners for large enterprises means that non-market allocation mechanisms can play an extensive role within such enterprises. These in turn, of course, further undermine the market mechanism by concentrating power in the hands of a few powerful monopolists.

The ability of markets to form a kind of back-up to other exchange mechanisms is extremely important. To achieve this, if markets and market relations are a widespread feature of the economic system, markets must be the principal exchange mechanism, supplemented and supported by indicative planning and other non-market mechanisms should the need arise. When these break down, a well-developed market mechanism is therefore present to fall back upon.

6

Market Socialism and the Reform of the Capitalist Economy

David Winter

THE long-term goal of any socialist government is to abolish the institutions of capitalism and create those of socialism. This chapter is concerned with the structural reforms that a market socialist government which comes to power in a capitalist economy should adopt to achieve this goal. Of course, such a government will also be engaged in a wide range of more orthodox policy interventions such as monetary and fiscal policy. It was argued in Chapter 5 that planning will have a role to play in improving the allocation of resources. This role complements, rather than replaces, market forces. Indicative planning is a means of co-ordinating the economic activity. Like monetary and fiscal policy, it is not suited to the kind of long-term reforms which will be considered here. Shorter-term policy will, of course, have to be informed by long-term considerations. The appropriate links between these different kinds of policy will emerge later.

Government intervention has traditionally played a large part in socialist political programmes. Socialists have naturally tended to favour 'big' government as the agency which would transform a capitalist economy into a socialist one. The Right has often criticized socialists on these grounds, claiming that big government tends to be inherently undemocratic and to be open to corruption and abuse. Rather than operating efficiently in the public interest, government bureaucrats, it is claimed, operate in their own self-interest by expanding their budgets and their influence. The Right claims that, in contrast to socialism, the great advantage of capitalism is that it can operate efficiently and spontaneously without government intervention.

This conveniently ignores the fact that capitalism also relies heavily upon government intervention. In the last fifty years every capitalist country has found it necessary, for various reasons, substantially to increase the size of the public sector. Attempts in the 1980s to 'roll back the state' by right-wing governments in the United States and in Britain have shown that, if such governments want to retain some degree of popularity with their electorate, they have to proceed slowly and cautiously. Even in Britain, where the public sector has been considerably reduced through a privatization programme, the power and inclination of the central government to intervene in the economy has not decreased substantially. In most countries where industrialization lagged behind Britain and the United States, such as France, Germany, or Japan, the state has always taken an active role in the development of capitalism. Even in Britain, the pioneer in many of these matters, the state played an important part.

In this chapter the focus of attention is shifted away from government intervention towards structural reform of the legal rules and regulations which govern the formation of economic institutions. Just as capitalist governments set the legal rules that provide a framework for capitalist institutions to operate, so a socialist government should do the same for socialist institutions. The main (but not the only) legal entity that supports capitalism is the limited liability company. This did not spring up spontaneously overnight. It is the creation of numerous acts of Parliament passed by governments which were trying to encourage the development of capitalism. If a socialist government wishes to change the nature of capitalist society, then the bulk of its efforts should be devoted to changing these and other laws which permit and encourage capitalist institutions to flourish. This kind of long-term reform is in stark contrast to the *ad hoc* interventions practised by the Labour governments of the 1960s and 1970s, when it sometimes appeared that their claim to power was that they could manage capitalist institutions more efficiently than right-wing political parties.

Since market socialism, like capitalism, favours markets as the predominant exchange mechanism, this poses the central question of how such an economy would differ from a capitalist one. If the chief objection to capitalism lies in the unequal distributions of income and wealth which are produced by the capitalist economy, is market socialism no more than an egalitarian version of capitalism? Is market socialism simply capitalism 'with a human face'?

To provide answers to these questions, it is necessary to re-examine the economic objections to capitalism and to identify the processes in capitalism which lead to outcomes in terms of the distribution of resources that socialists find unacceptable. The possibility of sustaining capitalism without relatively large dispersions of income and wealth needs to be considered. This will lead to a discussion of whether the socialist objections to capitalism are simply confined to the end states which capitalist institutions can be expected to produce, or embrace the means by which it arrives at these ends as well. These fundamental problems have, of course, been at the centre of discussion and debate on the Left for many years. In one chapter, I shall only be able to give a very brief outline of some of the more important arguments. Nevertheless it is of interest to see how these questions can be approached from the perspectives of market socialism, with its emphasis on decentralization rather than from the point of view of a socialism that places greater emphasis on centralized procedures.

Before considering structural reforms, it is, as I have said, necessary to re-examine the traditional socialist objections to the capitalist economic system. The argument identifies the structure of property rights as being of fundamental importance. While this echoes much nineteenth-century socialist thought, it is in contrast with more recent socialist writers (see, for instance, Crosland, 1964). First, a summary is given of some recent results in the theory of exploitation. Here the differential private ownership of the means of production is shown to play a key role in determining exploitation. This naturally leads on to a discussion of nationalization which has traditionally formed a major plank in socialist programmes. It is argued that national-ization is, in general, an inadequate policy to reform the capitalist economy.

Defenders of capitalism sometimes suggest that the inequalities which arise in capitalist economies are the result of particular historical circumstances instead of being an intrinsic feature of capitalism itself. In the following section I give a number of arguments which show that the dynamic processes of capitalism can perpetuate an unequal distribution of the means of production. This inequality will allow both the exploitation and the domination of workers to persist. After considering some libertarian arguments in favour of capitalist enterprises, it is suggested that individuals may own their own companies but that they should not be able to sell their ownership rights except on conditions favourable to the work-force.

The final part of this chapter discusses some of the problems posed by modern corporations. Since these separate control of the firm from the private ownership of assets, it is unlikely that they can be relied upon to produce efficient results. In addition the modern corporation is a centre of economic power which is often outside the control of national governments. The bureaucracies that control these organizations are well placed to abuse and to exploit their position. Both the inefficiencies and the abuse of economic power can be reduced, if not eliminated, by placing both ownership and control in the hands of the entire work-force. The concluding section summarizes these arguments and places them in the overall context of the policy goals of a market socialist government.

EXPLOITATION

Workers are exploited because capitalists control the means of production. This control is usually exercised by means of ownership rights. These can take the form of direct possession or of shared ownership institutions such as partnerships. Most commonly of all, capitalist firms are limited liability companies owned by numerous shareholders. In fact in modern capitalism, dominated by large corporations, the ownership of the means of production has become divorced from the control. The consequences of this are discussed below. In this section it will be assumed that capitalists control the firms which they own.

The analysis of exploitation under capitalism lies at the heart of Marxist thought. Workers are exploited in the sense that, when they sell their labour services to the capitalist, the value of these services to the capitalist is greater than the value of the goods and services that workers can buy with their wages. To formulate this argument, Marx used the labour theory of value, which rests on very restrictive assumptions. A restatement of the theory of exploitation using a less restrictive framework has been recently undertaken by John Roemer (1982 and 1986).

Roemer shows that under certain conditions there is a straightforward relationship between exploitation and the private ownership of the means of production. These conditions require that all workers can be regarded as being fundamentally the same in the amount of talents they possess. If this is the case, then those with a relatively smaller endowment of industrial assets will be exploited by those with larger holdings. Those who own large holdings of industrial assets will thus become capitalists in the sense that they will hire others to work their assets. Workers who own no industrial assets will sell their labour to capitalists. In between there are a variety of middle groups who may sell their labour, work their own assets, and hire others in differing combinations.

The point that Roemer makes is that, in this kind of world, those who own large holdings of industrial assets always exploit those who begin with less. The notion of exploitation under these conditions remains straightforward and precisely corresponds with the ownership status of the participants in the economy. Those who have only their labour skills to sell do so to those who own assets. Workers will be always be poorer than employers. The rich exploit the poor.

The analysis is not so simple in an economy where there are substantial differences between members of the society with regards to their inherited abilities and tastes. The comparatively straightforward relationships between wealth, ownership, and exploitation break down. It is possible to envisage skilled rich workers exploiting poor capitalists and even, under certain circumstances, poor workers exploiting the rich. These form counter-examples to the original Marxist position that under capitalism capitalists always exploit workers.

It is important to bear in mind that, while these counter-examples can be shown to be possible from a theoretical point of view, this does not imply that they are numerically important or indeed that they actually exist at all. Our intuition suggests that there may be some people who earn high incomes from their labour (surgeons, film stars, lawyers, etc.). These people may, as a result, be relatively privileged compared with a small shopkeeper, who, although he or she may technically qualify as a capitalist, nevertheless has to struggle to make ends meet. Roemer shows that from a theoretical point of view, in a world with a wide variation of skills and tastes, it is impossible to say who is exploiting whom—the rich lawyer or the poor shop-keeper. Nevertheless these inequalities, and the exploitation to which they may give rise, can be regarded as insignificant when compared to the large and common differences in income between those who earn average wages or less and have little or no property income, and the rich, the bulk of whose income is derived from unearned sources.

This rather theoretical argument serves to make one point clear. Exploitation in modern industrial capitalist economies arises from two kinds of inequality. The first and probably the most important is the result of the differential ownership of productive assets. The second arises from the fact that people bring different skills to the labour market.

Of course, differences in skills may simply be the result of different educational opportunities. Since a better education in capitalist societies can be bought by the rich, some of the observed differences in marketable talents is the result of inequalities in parental income. Nevertheless, even if such sources of inequality were somehow abolished, some differences in skills and thus in the market wage available to the possessors of these skills would undoubtedly persist.

A theoretical distinction can be made between those differences in income that arise from the inheritance of productive assets (and the associated acquisition of skills through parental purchases of additional education) and those that arise from innate differences in the talents with which people are born. Socialism is first and foremost concerned to eliminate the inequality that arises from differences in the ownership of industrial assets. The reduction in income inequality that results from differences in innate skills is clearly a distinct and difficult task. The obvious place to begin to address this problem is in education (see Chapter 4).

The traditional socialist method of eliminating exploitation was to take the means of production into public ownership. Nationalization has been the principal way of achieving this. In some countries, such as those in Eastern Europe, all large private holdings of assets were nationalized in a comparatively brief period, but such wholesale nationalization has never occurred under a democratic government following democratic procedures. Rather, various industries have been selected in turn for nationalization. This kind of policy is no longer fashionable, but it has played such an important role for socialist governments in the past that it is important to consider its strengths and weaknesses.

NATIONALIZATION

The first step in a gradualist policy of nationalizing the means of production involves identifying the order in which industries are to be nationalized. One criterion can be expressed in the traditional phrase, the 'commanding heights of the economy'. These would be the first priority in a nationalization programme. The question then arises as to where these heights can be found. It is probably difficult to realize now how much time used to be spent on the Left drawing up lists of which industries should be nationalized. Should steel be nationalized? What about the aircraft industry. How important is the machine-tool industry? And so on. Naturally one would expect that the most important sectors of the economy will change as new technologies are developed. Once it was coal, iron and steel, and the railways. Now it may be electricity, electronics, and banking. Industries that were nationalized decades ago may well have lost their importance. Should they remain nationalized or are they to be pensioned off to the private sector?

Having compiled the list of important industries, socialist governments are faced with the time-consuming and often politically costly process of passing the necessary legislation and settling appropriate compensation. If conducted on an industry-by-industry basis, this may take many years to complete. Once an industry has been nationalized, the task of controlling it in appropriate ways begin. This is at the heart of the problem with such a programme in practice. Enormous effort is taken to nationalize various industries, but the substantive issues of management and control after nationalization tend to be ignored or taken for granted.

It would be unfair and inaccurate to summarize the very extensive historical experience of nationalization in Britain and elsewhere as entirely consisting of failure. Some nationalized companies operate efficiently and effectively. But market socialists are not immediately attracted to the idea of organizing industries into large public monopolies. Governments find them difficult to control, their workers do not find that they provide enlightened and sympathetic employers, and their customers are often not excited by the standards of service and quality that they offer. It may not be possible in some circumstances to avoid nationalized public monopolies, but they should probably not be a principal feature of a market socialist society.

Socialist governments often pay a high price for their nationalization programmes and it is not obvious that they reap compensating rewards. The underlying problem with national-ization is that it almost always involves a separation of ownership from control. 'Ownership' is vested in the hands of the people, the working classes, or some other large collective. Control resides with a managerial bureaucracy that has little incentive to organize production efficiently or in ways that the workers find convivial. Those in control will tend to organize the industry in their own interests. These will not necessarily coincide with either consumers' or workers' interests. Given that they are also often legalized domestic monopolies, they are well placed to exploit their economic power.

Politicians are supposed to control nationalized industries in the interests of the electorate. Sometimes of course they succeed in doing so. But on the whole they are neither technically equipped nor do they have a strong incentive to become identified with the activities of the industries which they purport to control. In addition the managerial bureaucracy of a nationalized industry can form a powerful interest group. Relatively small in numbers, but well informed, armed with a strong incentive and with access to senior government circles, such a group is well placed to secure political favours that the more dispersed interests of the electorate at large may fail to secure. Political control of nationalized industries under such conditions cannot be relied upon to ensure that these industries perform in desirable ways, from the point of view either of their customers or of their workers.

The argument here is not that nationalization is never a wise socialist policy. There may well be particular circumstances or particular industries for which the best alternative is to take an industry or firm into public ownership. Nationalization is not being ruled out altogether. What is being argued is that, although nationalization does seem attractive to socialists in that it is a method of abolishing the private ownership of industrial assets and thereby the exploitation to which such forms of ownership give rise, it is not an attractive method of reforming the whole economy. The chief reason for this is that, just as with the modern corporation, the separation of ownership from control will lead to an industry operating in the interests of those who control it. This produces inefficient performance that serves the interests of neither workers nor consumers. In addition, socialist governments that embark on extensive nationalization programmes within democratic institutions find that the process is a time consuming and costly one. It distracts them from long-term policies which would probably be more fruitful.

THE DYNAMICS OF CAPITALISM

If it is granted that nationalization is too blunt an instrument adequately to reform a capitalist economy, let us return briefly to the arguments which led to its adoption as a major policy for socialism. Ignoring for the moment the theoretical complexities that I have already mentioned, capitalists exploit workers because, when they participate in economic activities, they can utilize the considerable advantages of large accumulations of privately owned assets. That is the essential premiss of the argument that leads to the conclusion that under capitalism workers are always exploited. But is this premiss correct? Is it always true that capitalism will be accompanied by the unequal distribution of assets? Advocates of 'popular capitalism' will answer these question negatively, implying that an egalitarian capitalism is a practical possibility.

In Europe, where capitalism replaced a largely feudal or peasant economy, the distribution of wealth was already very unequal. In this context the rise of capitalism in conjunction with the industrial revolution may have had, in the end, egalitarian consequences in relation to the preceding economic system. This is a subject that has been frequently discussed by economic historians (see, for instance, Hobsbawm, 1968).

The actual historical development of capitalism, however, is not what is at issue. What is important to understand is the principles which govern the performance of a capitalist economy. If the source of exploitation is the unequal distribution of capital, the question naturally arises as to whether it is possible to envisage a capitalist system that would have an egalitarian distribution of capital. If so, such a system would not be exploitative. To pose the question in another way, could capitalism not have been an economic system in which a few rich workers exploited numerous poor capitalists?

As we know, it has not tended to work out like this, and the reasons are worth considering. Everyone is endowed with some labour power. There may be differences in inherited skills which will give rise to income inequalities, but these inequalities are probably extremely small in comparison with those that can arise from the fact that the majority own no assets whatsoever, while a small minority own huge accumulations.

A persistent feature of capitalism is the relative scarcity of the physical means of production and the relative abundance of labour. This abundance has been artificially stimulated by dispossessing people of their entitlements to productive assets. In Britain the enclosure of land effectively dispossessed a large part of the population, many of whom became poorly paid industrial workers for the new capitalist class. Nevertheless there are two reasons why we would expect capitalism in its early stages to be accompanied by relatively high returns to capital and correspondingly low returns to labour. First capitalism usually involves more capital intensive forms of production than the agriculturally based systems that preceded it. Capital by its nature starts off in short supply. Market valuations of scarce inputs naturally involve a relatively high return. A corollary of this argument is that, as investment increases the volume of capital in a market economy, the return to capital can be expected to fall.

This is not the Marxist 'Tendency for the Rate of Profit to fall'. It is a simpler process by which the return to a relatively scarce input in production will be relatively high. As capital becomes less scarce, we would expect this return to fall. A second reason for the relatively high return to capital has been the high growth rates of population that were historically associated with the introduction of capitalism and industrialization in Europe in the nineteenth century and in the Third World today. Assuming that capital and labour serve as potential substitutes for each other in the production process, if the supply of labour grows, its price relative to capital will fall.

Shortages of capital and a rapidly growing labour force have usually been present when capitalist institutions started to take root. This gave rise to high returns to capital and correspondingly low returns to labour. However, these tendencies are no longer present in most European capitalist countries. The original acute shortage of capital has long since disappeared. The rapid growth in populations has also stopped. The numbers of people of working age can be expected to stay roughly constant or in some cases actually decline. This should bring about a relative increase in wages and a relative fall in the returns to capital.

Some argue that this has already happened (Samuelson and Nordhaus, 1980). A comparison of the share of wages in total income in most European economies at the end of the nineteenth century with the share now indicates that these forces seem to have been at work. However, there are good reasons for believing that capitalism will not converge to an egalitarian income distribution as investment makes capital more abundant and, at the same time, the number of potential workers in the population remains roughly constant.

The uncertainties surrounding the generation of profits tend to ensure that successfully surviving owners of capital are relatively rich. Profits are an uncertain residual payment which can be either positive or negative. Not all capitalist firms are successful. When a firm makes persistent losses, the value of its assets will fall. Either such a firm will be taken over, and its assets managed more profitably, or its assets are scrapped completely and disappear from sight. So, whereas poor workers 'are always with us', until they die the sooner as a consequence of their poverty, poor capitalists are hardly ever to be seen, except briefly during a period of bankruptcy.

The failed capitalist 'disappears' and becomes a member of the majority of society who own no productive assets, and can sell only their labour skills. Successful capitalists, on the other hand, can diversify their risks, exploit economies of scale, and accumulate monopoly power. This serves to exclude the entry of those who own and control smaller quantities of physical capital. This process maintains and increases the concentration of the ownership and control of physical assets, giving the owners increasing power in relation to those who possess only labour skills. Such concentrations of power may be undermined by the forces of competition. But successful competition usually can only come from other large firms or through technical innovation.

Technical change plays a crucial role in the dynamics of the capitalist economy. Before a new invention can be put into practice, the innovating capitalist has to invest in new kinds of capital equipment. There will usually be greater uncertainties surrounding such ventures than in investment projects that use more conventional technologies. It is not surprising that sometimes long delays can occur before new inventions are applied to productive processes. In Chapter 5 it was argued that these uncertainties may lead private decision-makers to under-invest. Here the important point is that new technologies often provide capitalists with opportunities to make investments with relatively high rates of return.

Once the uncertainties associated with innovation have been overcome, a successfully innovating capitalist can reap large profits. At the same time those firms which rely on older technologies will find their capital equipment less profitable. They may respond by trying to reduce wages or by forcing their workers to work harder. Their ability to do this effectively will largely depend on conditions in the market for labour and the legal arrangements that govern its employment. However, the important consequence of the new technology is to make older capital less profitable and, often, redundant. In such cases, unprofitable capital may have to be scrapped earlier than planned.

This process can be regarded as one in which new technology creates new shortages of capital. By making old capital equipment unprofitable, it can help to create scarcities of new kinds of capital. Of course, one technical innovation will only do this once. But what we observe in capitalism is the continuing introduction of new technologies and the continuing creation of new relative scarcities of capital that embodies new technological processes, and the writing-off of old unprofitable capital. This ensures that the return to capital remains relatively high. It prevents the long-run decline in these returns as the capital of economy grows over time.

This argument suggests one reason why there is a long socialist tradition that seeks to limit the introduction of new technology. New innovations can weaken the position of labour relative to capital. In addition innovative firms may be located in different regions (or countries) from those where the older firms are situated. British readers will be well aware of the socially disruptive effects that can follow the closing down of industries that are no longer profitable.

It is important to realize that, although it may be desirable on communitarian, social, or other grounds to limit new innovations, this runs counter to the encouragement by competitive markets of profitable production. The socialist reaction to this process should not be to prevent innovation but to ensure that the profits from such innovations do not exclusively accrue to the capitalists who own the new capital. Workers who use the new production processes will have valid claims to it as well (see below). An indicative planning process will provide a forum where the adverse social consequences of certain kinds of investment can be considered before the investment takes place.

So capitalists can preserve the position of capital relative to labour, by creating new shortages of capital through the innovation process; an alternative method is to 'de-skill' labour. In market economies, workers can usually expect to increase their wages by acquiring new skills that are in comparatively short supply. As a result, the ever-increasing skills of the work-force, it may be supposed, will reduce the inequality between workers and capitalists.

A feature of modern capitalist innovation is that it gives capitalists an opportunity to try and arrest this process. Capitalists usually find that it is more profitable to introduce new capital which requires low levels of skill on the part of workers. Skilled workers are sometimes deliberately excluded from working under such conditions, even for unskilled wages. Braverman (1974) in an influential book has argued that capitalists require workers to exercise minimal skills in their work. Instead of employing skilled workers who can take an interest and a pride in their work, workers are reduced to being an input into the production process. They are in effect treated as if they were another form of material input.

Employers find that unskilled workers are easier to control. As a result it is easier to ensure that they work in the required manner. In addition the domination of an unskilled work-force is easier to sustain (see below). Of course the tendency for capitalists effectively to reduce the demand for certain kinds of skills and replace those workers with the unskilled can have complicated effects over the whole economy. It may mean that those with skills which are still in demand can increase their relative wages as a result. It is possible that the unskilled wage may also rise, in the unlikely event that the supply of unskilled workers is limited. The overall consequences will again depend on conditions in the labour market as a whole. Nevertheless it is noteworthy that capitalists find that 'de-skilling' is a valuable way of ensuring that their profit rates are maintained. This, at the very least, does not encourage the belief that such a system will inevitably become egalitarian.

I have argued that the dynamic processes of capitalism make it most unlikely, if not impossible, for a capitalist system to converge to an egalitarian distribution of income and wealth. As a result exploitation can be regarded as a persistent feature of capitalism. In practice, exploitation of the kind I have been discussing is a rather abstract concept. For instance, no one can identify who is exploiting whom, especially given the complexities of modern corporate capitalism. A more obvious and direct consequence of the control exercised by capitalists is the domination of workers while they are at work.

The forces of competition ensure that profit margins are always being squeezed. It is in the interests of capitalists to reduce wages and force their workers to work harder. The relative scarcities of capital and labour ensure that in a market system workers will, on the whole, be paid relatively little. This means that capitalists have the power to arrange and control the lives of workers at the workplace. The harsh conditions of work can severely impoverish workers' lives. This will be aggravated by their relative poverty. It deprives them of the inclination and ultimately of the ability to be resourceful and creative in their labour. The resulting deprivation and suffering continue to form a feature of capitalism that for many constitutes its chief iniquity.

The kind of working conditions associated with mass production and the production line are a striking example of this kind of domination. It is sometimes supposed that production-line techniques are being phased out in a 'post-industrial' economy. This may be true, but it is unclear that dominated work activity is being phased out even in the most advanced economies. New forms of domination can appear and, in the developing world, the abuses of production-line methods show few signs of disappearing. This is an important reason why market socialists favour co-operative methods of production (see Chapter 7).

One way to inhibit the concentration of the ownership of capital is the system of poll grants discussed in Chapter 8. Under these, everyone would receive an equal allocation of capital in their youth. This would enable them to begin accumulating capital, and an element of equality among capitalists would have been achieved. This kind of scheme has definite attractions, but these will probably be of short-term value if it were put into practice in the context of capitalist institutions. The unsuccessful would eventually lose their ownership rights. The successful, if they were able to trade in shares, would find, by diversification and takeovers, that they would be able to acquire large holdings of assets. The successful would reach positions of power that would enable them to exploit and dominate other workers.

It is sometimes argued that there is nothing wrong with capitalist acts between consenting adults (Nozick, 1974). Provided that the capitalist has acquired the productive assets legitimately, and here I would rule out inheritance, why should the inventors of ideas or those with entrepreneurial talents not use them to form privately owned companies, hiring workers at the going wage and acquiring the profits from the success or the losses from the failure of such an enterprise?

Some economists have thought of the innovating capitalist in heroic terms—individuals who by their sole efforts build large companies which give thousands of workers prosperous employment and millions of customers the benefits of their inventions. People like Andrew Carnegie, Henry Ford, or William Morris from the early years of the century come to mind; Rupert Murdoch and Alan Sugar from more recent times. But while market socialists would encourage small-scale enterprises, they will wish to restrict or to abolish large-scale privately owned capitalist firms.

As companies grow, their assets also tend to grow in value. The control of these assets should not be confined to a small number of owners. The use and development of the productive assets of any society have great importance for both workers and consumers. A capitalist is not entitled to sole control of the assets that are accumulated thereafter, simply because he or she started the firm with a bold idea or a good invention. Without the contribution of workers hired by the capitalist, the capitalist would not have been able to accumulate assets in the first place. This in itself should entitle the workers to a share of those assets. It is remarkable how self-made millionaires consider that their achievements were almost solely due to their own efforts, whereas in fact they were the consequence of the efforts of thousands of others as well as their own.

It is sometimes argued that, since the workers were paid the going wage, no further reward is required. They voluntarily engaged in a capitalist act, and deserve no share of any profits that are subsequently made. To see why this argument is wrong, consider the case of slavery. Is there any reason to ban acts of enslavement between consenting adults? On the face of it not, and it is indeed probable that there are many people in the world who would voluntarily enslave themselves if as a result they received a guaranteed standard of living. But most people would not support such an institution and with good reason. On moral grounds, we believe people should not be owned by others. Similarly socialists do not believe that labour time should be traded in an identical fashion to commodities. People who work in a team, in however humble a capacity, will have devoted some of their life to the success of whatever tasks are being undertaken. This should give them an entitlement, however small, to any profits that ensue.

An additional argument that supports the sole retention of profits by capitalists is that they take on sole responsibility for the risks involved. Profit, if it occurs, is a reward for undertaking these risks. Workers face no risk and so deserve no share of the profits. The objection to this argument is that it is quite wrong to assume that the workers undertake no risks. If the firm is not successful, workers will be laid off. Even in an economy with full employment, this will involve redundant workers in considerable costs. The situation will be worse if the firm actually goes bankrupt. In this case, workers can lose wages that are owed to them as well as their jobs. Most workers therefore do undertake some kind of risk when they join a firm, while capitalists who own a variety of productive assets can reduce the risks they face by diversifying their operations. If they are very successful in doing this, they may face very little risk at all.

This is not to say that the founders of a firm may not bear more risk than its workers, nor that those who have good ideas or make productive inventions should not be given considerable compensation for doing so. A market socialist would not want to impose a uniform distribution of wages on the economy. But the argument does imply that the sole ownership rights of capitalists should be abolished when the firm reaches a certain size.

It is possible that, although capitalists do not bear all the risks involved, they may nevertheless be forced to bear too much risk in a traditional capitalist system. It might be more efficient if there were other ways by which innovators could reduce their exposure to the possibility of unfavourable outcomes. Indicative planning may serve to reduce the risks involved in making investments. The exchange of information can reduce uncertainties about future market conditions. Innovators may then feel freer to use their talents for innovation. Capitalists may not be the sole bearers of risk, but the concentration of risk upon them may be unproductive in terms of making new investments and starting new companies.

It may be suggested that the great innovators would not innovate without the possibility of acquiring massive wealth and power as a consequence. However, most people with exceptional talents take pleasure in exercising them. In a market socialist society they would find that their talents would provide them with relatively high wages. Many capitalist firms today hire skilled scientists to undertake inventions for them. It is not self-evident that such firms are slow to make new discoveries. Innovators under market socialism would be denied the power that accrues from the ownership of great wealth. If there are innovators who will not innovate without such incentives, they will have to make their innovations elsewhere. The external costs that they impose on the rest of society are too high.

The costs of ensuring that no capitalist firms exist at all would obviously be absurdly high. Capitalist firms that are sufficiently small do not pose a serious threat to the well-being of others. If labour market conditions are reasonably favourable, workers will be able to leave if they find conditions unacceptable. At a certain size this is no longer true. It therefore seems reasonable to require capitalists to share ownership with the work-force once their firm reaches a certain size. An attractive solution to the problem of how large a company should be before it ceases to be privately owned is to allow the work-force to make the choice. Companies that are more efficient when run in a capitalist manner could then remain capitalist so long as the capitalist concerned could persuade the work-force to continue with this form of enterprise. An obvious difficulty with this scheme is that new workers would be under pressure to conform to the capitalist's wishes. The capitalist might pay his old workers comparatively high wages to prevent them from forming a co-operative. This kind of firm would then have no incentive to expand as it should.

The alternative method of imposing a limit to the size of the capitalist firm is to prohibit the sale of shares in the firm's assets, except under conditions where the work-force can take a majority stake.[1] The owners would be able to build up the firm to any size so long as they did not attempt to diversify ownership by using the legal means of limited liability. The sale of partnerships and firms owned by single individuals would also only be allowed under conditions where the work-force could buy out the assets. Fiscal incentives would encourage these sales of assets to the work-force.

It may be objected that under such arrangements large capitalist firms will still persist. However, they will only be able to do so in the life span of the founders. Most founding capitalists will wish to share the ownership of the firm, and, if possible, to realize the expected future profits of the firm. This may occur when the firm reaches a certain size or it may be at the retirement of the founders. At this point it is desirable to

---

[1] I am indebted to Saul Estrin for this point.

prevent the continuation and expansion of capitalist relations within the firm. Further growth should be under the active participation and control of the work-force. For rather different approaches to these problems, see Nove (1983) and Ryan (1984).

CORPORATE CAPITALISM

So far I have been considering capitalist companies that are controlled by their owners. Strict inheritance taxes in a market socialist society will ensure that these firms will have been started within the lifetime of their owners. If large firms emerge, it will be the result of the efforts of 'heroic capitalists'. Such figures would be well rewarded within a market socialist society, in terms of a high income and probably in terms of respect and status too. But they would not acquire permanent ownership rights over large quantities of productive capital.

Most owners of capital in modern industrial economies are not, of course, heroic at all. As capitalist firms have expanded, so has their share ownership. The world-wide integration of stock exchanges continues a process by which ownership and control of large companies are increasingly separate. Most large companies are owned by a large number of small shareholders and by a number of financial institutions such as pension funds, investment banks, and insurance companies. I shall consider these two different classes of owner in turn.

Small private shareholders do not in themselves constitute a threat to the egalitarian principles of market socialism. On the assumption that absolutely no wealth will be inherited, if some people wish to hold their savings in the form of shares then this would present no difficulties. In fact, the main problem derives from the very lack of control such shareholders exercise over the assets they own. To expect small shareholders in a large company, such as British Telecom, to make informed decisions about the use of corporate assets is absurd, especially if they live hundreds or even thousands of miles from where the firm operates.

Usually, in a large modern corporation, control will pass to the management. Provided profits are maintained, the management is free to conduct the affairs of the company as it wishes until a takeover bid is made by a rival company. A public, but not always well-informed discussion takes place concerning the alternative plans put forward by the rival management groups. The shareholders make their decision and then find themselves locked into the victorious management team. Few economists have any faith that this process leads to the efficient use of a corporation's assets.

Small shareholders seek to increase their savings through dividend payments and capital gains. Some may enjoy the pleasures of gambling on risky assets. They are not particularly interested in or informed about the complex questions of managing a large industrial company. For this reason they cannot be expected to be owners of assets who seek to improve and maintain their value in socially desirable ways. They will be the inevitable victims of the present management or of rival managements who will have access to much, if not all, of the information needed to make the relevant decisions. The legitimate interests of small shareholders can be perfectly well secured without owning shares in large corporations. By using options and bets, savers can construct portfolios which will provide them with future income at varying amounts of risk. Small shareholders do not want to own, in any active sense, industrial assets. They should be given more attractive alternatives.

Institutional investors present similar problems in a different form. In conjunction with the spread of a world-wide system of linked capital markets, there has recently emerged a new class of investment advisors and financial analysts who manage the portfolios of institutional investors. Often quite young, these market operators are free to speculate with huge accumulations of financial wealth. It is too early to be certain about the consequences of this new concentration of financial power, but the current indications are not encouraging.

In Britain, institutional investors have traditionally not played an active part in the management of the companies that they, in practice, own. Again they have no inclination or particular talent to do so. Insurance companies, pension funds, etc., all own shares in large corporations for exactly the same reasons as do small shareholders—to earn dividends and to make capital gains. They may be in a position to take a longer-term view and the size of their portfolios may enable them to take on riskier investments, but they are just as uninterested in and as ill-equipped to deal with problems of controlling industrial assets of great value to the whole society as the small shareholder.

In practice, by delegating their investment choices to market operators, they may act more speculatively and more sensitively to short-term fluctuations than the small shareholder. The highly paid investment advisor will wish to show the value of his or her advice in the short term, rather than over the much longer-time horizons that are appropriate to pension or insurance investments. This kind of short-term bias in the investment decisions of large and powerful financial institutions can be seriously detrimental to the long-term investment prospects of the economy. It is not the purpose of this chapter to provide the appropriate policies to deal with this issue. The Labour Party has made proposals in this area. The point of the argument is that the modern corporation, set up as a limited liability company, owned by shareholders who trade these shares on stock markets around the world, and controlled by a largely independent management, is a powerful and probably also an inefficient institution.

Oliver Williamson, one of the few economists who has seriously studied large corporations, has argued that by economizing on the costs of making transactions, the modern corporation can exploit 'organizational' economies of scale. He uses this argument to suggest that the modern corporation is an efficient organizational form (Williamson, 1986). While this argument may be correct, it still does not imply that the ownership structure of the modern corporation is efficient. Managers may be able to exploit organizational advantages but will do so in their own interests, not necessarily in the interests of owners or workers. Estrin in Chapter 7 suggests a way of organizing co-operatives which has the possibility of preserving the divisional form of the modern corporation—and thereby preserving any organizational economies as well—while completely changing the ownership structure.

Ownership rights should be given to those who have an incentive and the information efficiently to maintain and to improve the productivity of industrial assets. One approach to the problem is to see that the modern corporation, by giving control to management, has begun to evolve into a kind of co-operative—one that market socialists would not wish to endorse, because of the division that still exists between workers and management, but a co-operative nevertheless. The division between workers and management, although important in practice, is often a left-over from an older, traditional form of capitalism. Workers almost certainly have access to valuable information that will be relevant to management decision-making. To divide the work-force from management simply because the management wishes to preserve its privileges is neither equitable nor efficient. As modern Japanese companies have realized, the division between management and workers often has no validity in terms of operating the company efficiently. As I have argued above, anyone who participates in a production process is entitled to a share of the profits. If it makes sense to give management ownership rights because they are the group who control a large modern corporation, then it also makes sense to extend these rights to all the workers involved.

CONCLUSION

The argument of this chapter has been that socialist governments wish to reform the capitalist economy because under capitalism workers are exploited and dominated by those who privately own the means of production. Nationalization of those means of production is not an attractive policy to pursue in general, although there may be circumstances when it appears to be the best alternative.

The unequal distribution of private wealth is an essential feature of the capitalist economy, not an historical accident. The dynamics of capitalism will lead to concentrations of wealth that will empower their owners to abuse those less fortunate. Radical capital transfer taxes may assist to prevent such accumulations from being passed on to descendants, but capitalist firms may nevertheless pose a threat to the egalitarian aims of socialism. However, to rule out small private firms altogether seems unnecessarily restrictive. A compromise is suggested whereby private ownership is tolerated so long as the owners do not wish to sell their assets. Should they wish to do so, and small-scale capitalists often seek to realize the future profits of the firms which they have created, the work-force should be encouraged to take over control of the firm. As Estrin suggests in Chapter 7, it may be more efficient to vest owner-ship in these cases with a holding company rather than with the workers themselves.

The analysis of the modern corporation suggested that the institutional ingenuity of capitalism in devising new forms of efficient organizations to suit changing circumstances has at last begun to fade. The inefficiencies caused by the separation of ownership from control can obviously be avoided by recombining them in new forms. If workers collectively control the assets with which they work they are less likely to be dominated in the workplace. The case for co-operatives is discussed in detail in Chapter 7. Labour–capital partnerships may also have a role (see Chapter 4). The structure of property rights will have to be changed to give workers control over production, and to prevent accumulations of wealth that in a market system threaten the welfare of others. In conjunction with such a reform, the ways in which workers acquire skills and the market in which they trade them will also be an important priority for a market socialist government.

The purpose of this chapter has been to re-examine traditional socialist arguments against capitalism. Market socialists, like the vast majority of socialists before them, wish to eradicate as completely as possible inequalities of income that arise from inequalities in the ownership of industrial assets. From this it follows that a socialist economy will have to abandon the private ownership of the means of production as the principal form of ownership in the economy. Strict inheritance taxes and a restriction on the size of companies that are privately owned are also necessary. Market socialists would also advocate the use of a market to allocate labour skills, though it can be expected to give rise to income inequalities. It is important that the labour market works efficiently and equitably as well.

It is not precisely known how large are the differences in inherited endowments of skills and talents. It is probable that, even if everyone had equal educational opportunities, there would still be some variation in the resulting distribution of skills and talents. Those with scarce skills, whose labour power would be valued comparatively highly by market processes, would receive a relatively high wage. Those who were unskilled and who tend to make up a large, but at the moment decreasing, part of the employed labour force, would find that their skills were valued less.

It should be clear that full employment becomes an essential priority for a market socialist government. This is not only because it increases the welfare of workers; it also plays an important role in the logic of the economic system being proposed. It makes little sense to advocate an equality based on the possession by all of some kind of marketable labour skills, if some workers find that there is no market in the skills which they happen to have obtained through the educational system. To achieve full employment, the government will have to arrive at an appropriate blend of micro-economic policies designed to improve the efficiency of labour markets and support retraining and migration where necessary. At the macro level, policy will have to aim for full employment without inflation. If some economically powerful groups emerge who can effectively ignore and override their own budget constraints, an incomes policy may be necessary. The preference of market socialists for small competitive units means that the need for an incomes policy may be less likely to occur than when large monopolistic structures dominate the economy.

Capitalism requires governments to construct the legal framework which enables capitalism to thrive. Similarly a market socialist government should devote most of its reforming efforts to replacing laws that define capitalist property rights with a socialist framework. The result should support and sustain the forms of ownership and the kinds of property rights that, it has been argued, are necessary for socialism. Once these fundamental legal reforms are carried out, a large government bureaucracy—which has been a typical feature of socialism in the past—will not be required.

7

Workers' Co-operatives: Their Merits and their Limitations

Saul Estrin

THE shift of emphasis on the Left from plan to market has brought to the fore the issue of how best to organize enterprises in a socialist society. There is no real problem for the committed planner. Publicly owned corporations should be managed by their managers, whose job it is to implement the plan. For 'left' libertarians, socialism is about equal entitlement to the means of production, with the question of how people choose to use their endowments in the production process left open. Indeed, the Croslandite view is that capitalist firms are an acceptable component of a socialist economy, provided that taxes and subsidies exist to eliminate inequalities. But there is also a longstanding socialist tradition which argues that fundamental changes in society must be intimately bound up with changes in the way that work itself is organized. This points to workers' self-management of industry. Since effective workers' control is contingent upon the enterprise itself, rather than, say, a local or central planning office being the basic unit of economic decision-making, market socialism is a particularly convenient framework into which the decentralization of production decisions to workers can be embedded.

Co-operatives are very much the flavour of the month on the British Left. This is largely for ideological reasons. Socialist authorities have been heavily involved in co-operative formation via local Co-operative Development Agencies (CDAs) as part

Many of the ideas in this paper were developed from a project on nationalization and privatization in France and Britain, financed by the Leverhulme Trust. I would like to thank Virginie Pérotin for comments and discussions.

of their broader economic strategy. Co-operatives are seen as a potentially successful organizational form in which socialist ideals do not necessarily conflict with commercial viability. We have seen an enormous upsurge in the number of such organizations in the United Kingdom, from less than twenty in 1975 to 330 in 1980, 1,400 in 1985, and perhaps as many as 1,600 today. The numbers working in co-operatives have risen from less than 2,000 in 1970 to more than 10,000 now. Recent empirical work suggests that the failure rate of co-operatives may be less than that of other types of small businesses, between 6 per cent and 11 per cent per year, and is certainly no greater (see Estrin and Pérotin, 1987).

The majority of British co-operatives meet socialist objectives, as conventionally defined, in that they have often been formed to produce for 'social needs' rather than purely for profit. Their activities are concentrated in the service sector (52 per cent of the total in 1980), particularly in restaurants, bookshops, other retail outlets, printing, house decoration, and record, film, and music-making. Such co-operatives frequently try to satisfy demand from the local community or other co-operatives rather than the wider market. They are also typically very small, with the average number of workers in each being around five in 1984. There is often an anti-growth, and indeed sometimes an anti-capitalist ethic, with the consequence that there are almost no large co-operatives in the United Kingdom. In 1981 only fifty co-operatives had a turnover in excess of £100,000, and even now none employs more than a thousand workers. This may also be because the bulk of co-operatives are new; their average age according to a 1983 Greater London Enterprise Board survey was five and a half years. If co-operatives set up before 1945 are excluded, the average age falls to three years. Whatever the reason, their small size and artisanal nature offer their members the attractive possibility of democratic control over their workplace, a welcome alternative for many to the hierarchical structure inherent in the capitalist corporation.

Co-operatives are also attractive within a campaigning socialist economic policy because they create jobs for dis- advantaged groups in society: the unemployed, women, blacks, and other ethnic minorities. A significant minority of co-operatives have actually been created out of the defensive actions by workers to preserve jobs following the bankruptcy of their capitalist employer, including the three well-known co-operatives supported by the then Industry Minister, Tony Benn, at Meriden, the Scottish Daily News, and Kirkby. The majority, however, have been formed as co-operatives from scratch, often with local government encouragement and support, as part of a positive effort by groups of workers to create businesses for themselves in areas of high unemployment, and for minority groups to create economic organizations sufficiently flexible to satisfy their aspirations.

It should be stressed that the British experience with regard to co-operative size and nature is far from typical. The Yugoslav system of economy-wide workers' self-management has excited interest for many years, and will be discussed in more detail below. Moreover, several Western European countries have large and vibrant sectors, in particular Italy, France, and Spain. The Italian co-operative sector is by far the largest in the Western world, with around twelve thousand co-operatives employing some half a million people. The bulk of the co-operatives are concentrated in the north and centre of the country, in Emilia-Romagna, Lombardy, and Tuscany, and operate in construction and services. In their evolution, Italian co-operatives have benefited from strong central organizations, public work contracts from local authorities, and from combines of co-operatives formed to deal with finance, design, marketing, and so forth (see Estrin, 1985). Italian co-operatives are therefore typically large (with an average of over three hundred workers in the top 10 per cent of firms), well organized, and supported by municipal authorities and the broader co-operative movement.

The French co-operative sector is rather smaller, with around 40,000 workers in some 1,200 co-operatives, but also well established and with a long tradition of production in printing, construction, and various branches of engineering. In recent years French co-operatives have also begun to emerge on the British pattern in services and consultancy, with relatively few workers and concerned with social rather than purely commercial objectives. A major difference, which no doubt contributes to the different sizes of the co-operative sectors in the two countries, is that over the past seventy years the French have developed a sound legislature and tax framework for the development of producer co-operatives.

In Spain, in addition to many small-scale artisanal co-operatives, there is the large federated group organized around the Caja Laboral Populaire in Mondragon. From their foundation in the mid-1950s, Mondragon co-operatives grew to employ some 8,500 workers in 40 co-operatives in 1970, and more than double that by 1983 (see Estrin, 1985). Mondragon co-operatives are concentrated in industrial manufacture, and successfully compete on both Spanish and world markets. Finally, there is a long tradition of producer co-operatives in the United States dating back to Robert Owen's Utopian communities. The bulk of co-operatives are clustered in plywood manufacture, but there has been a recent upsurge in services paralleling European developments (see Jackall and Levin, 1984).

This apparently gratifying combination of economic viability and ideological acceptability has led many to see worker co-operatives as an important precursor of the organizational form appropriate for a socialist society. Such a view has struck a chord amongst those who, recoiling from the Soviet-type system, have pointed to the Yugoslav experiments with social ownership, markets, and workers' self-management which we detail below (see Comisso, 1979). But the question remains whether these organizations really represent a blueprint of how to run firms in a market socialist future. It is to this issue, and in particular to the insights we can gain from economic theory, that this chapter is devoted.

In the next two sections I summarize the principal merits of co-operatives from a socialist perspective and outline the various ways that such organizations operate in practice. The problems which economists have suggested co-operatives will face in the market-place are the subject of the following section. Despite the pessimistic implications, I go on to propose institutional arrangements and legal changes which could ensure the efficiency of a market economy in which producer co-operatives are the predominant form of enterprise. The final section is concerned with the transition to these new arrangements.

WHAT CO-OPERATIVES CAN OFFER

The idea that people should own and control their own firms, rather than work for capitalists, has been around since the industrial revolution, and has spawned a long, if somewhat marginal, tradition in socialist thought. The notion was associated with Utopian thinkers such as Owen, St Simon, and Fourier, and was first developed formally in Paris by Buchez during the 1830s and 1840s. He proposed the formation of 'working mens' associations', in which control by capital was replaced by workers' self-management and group ownership of the means of production. Although several hundred producer co-operatives were formed in both Britain and France during the latter part of the nineteenth century, under Marxian influence the dominant strand in the labour movement gradually became a concern with public ownership of the means of production. By the early twentieth century, socialists such as Beatrice Webb were highly dismissive of producer co-operatives (see Webb and Webb, 1920). The exception, of course, was G. D. H. Cole, with his endorsement of a British system of workers' self-management—Guild Socialism.

The attraction of firms which are owned and run by their workers are easy to see. First, some would argue that enterprises in which capital hires labour breed exploitation (see Chapter 6). In contrast, when labour hires capital, the means of production are finally put in their place, as a tool of labour power rather than its master. This perspective harks back to a vision of pre-industrial days, when the role of capital in the production process was less and when artisans could perhaps hope to finance the equipment that they needed in order to retain control over their working lives. As production processes have become more capital intensive and economies of scale have increased the least cost size of plant, the argument goes, workers have had to hand over their rights to self-determination in return for access to the means of production. Moreover, the concentration of control and ownership in the hands of capitalists has allowed profits to be maintained at the expense of wages. The solution is for groups of workers to form productive organizations specifically geared to upholding the rights of labour as well as the supply of output to the market-place.

A more recent strand of the literature would stress the importance of democracy in the workplace. A starting-point would be the sharp contrast between democracy in the political process and autocracy at peoples' places of work. Democratic involvement in political decisions is now regarded by most as a fundamental right for all adults. Yet this sits uneasily with the fact that hierarchical systems of control, paralleling dictatorship in the political arena, are taken for granted in the enterprise. If we regard involvement in decisions which affect our lives as a basic prerequisite of a humane society, the democratic processes which govern our political life must be extended into the workplace. This will act to diffuse power, by giving people an equal say in enterprise decision-making. If anything, the current arrangements in the enterprise tend instead to undermine political democracy, by devaluing the potential contribution of the people at the bottom of the hierarchy and restricting their political skills. In contrast, by increasing rights at the workplace and giving experience in decision-making, workers' self-management could help to buttress political democracy.

One of the key problems stressed by observers of the capitalist enterprise is the dissatisfaction or alienation felt by a significant proportion of the work-force. Workers have no say in the major decisions affecting their working lives: the production processes used, the pace of manufacture, the noise levels, manning arrangements, the layout of the plant, and the decision to increase or reduce the labour force or even to close the factory. Their dissatisfaction comes out in a number of ways: their attitude to work, to their supervisors, to management in general, and to the owners is often highly negative—the 'them and us' syndrome. If the labour force is not unionized, this often leads to uncooperative attitudes, inflexibility with regard to work practices, and high rates of absenteeism, shirking, and labour turnover. In a unionized environment, the dissatisfaction also makes itself felt through unions' militancy, industrial action, and strikes. Once again many of those problems are in principle soluble in a system of workers' self-management. Individual employees are given an equal vote in determining all aspects of company policy, which should help engender a new attitude of mutual support and co-operation in the workplace. Moreover, since each worker now has a stake in the profit of the firm, material incentives can act to reinforce a fundamental change in attitude towards work. The consequence of this reduced alienation and increased involvement may therefore be substantial gains in the productive efficiency of the organization.

A final major grievance with the traditional capitalist system is the persistence of inequalities in the distribution of income. These in large part arise from the allocation of corporate profits as dividends to a small number of owners, and from the payment of significantly higher wages to people with greater skills, in particular to managers and professional experts (see Chapters 4 and 6). In a system of producer co-operatives, non-retained profits are instead distributed to the work-force, a significantly more dispersed group than shareholders in most cases. Although inequalities may remain between workers in sectors of low and high profitability, the replacement at the economy-wide level of a small owning group by the labour force as a whole will act to equalize the distribution of income. In addition, the distribution of income between people of different skills within each enterprise becomes a matter for internal debate and vote under self-management, rather than being imposed from above by management. While the outcome will still reflect to some extent the market position of those with special skills, it is likely to be more egalitarian than pertains in capitalist firms. In particular, it seems unlikely that the very high salaries and other perks accorded to themselves by top managers would survive open scrutiny and democratic vote by other employees. For example, in Mondragon the maximum pay differential from top to bottom has been determined democratically at 3:1. Co-operatives in Italy and France are similarly egalitarian.

THE VARIOUS WAYS TO ORGANIZE CO-OPERATIVES

These arguments in favour of workers' self-management tell us little about the best way to organize a producer co-operative. Moreover, they stand in stark contrast to a now abundant literature, both theoretical and empirical, which argues that such organizations have serious deficiencies as the basic unit for organizing production in a market system. Since many of these criticisms derive from the way co-operatives are organized, it will be useful to survey the institutional alternatives. This will have considerable relevance for the argument that follows, because we will find that producer co-operatives, as conventionally organized, *do* suffer from a number of serious flaws. However, these arise from deficiencies in the way that co-operatives are organized, rather than more profound drawbacks to co-operatives as an institution. We are, therefore, able in the final section to propose arrangements which could support an efficient producer co-operative sector.

The central issue for the workers' co-operative is one of ownership, with its ramifications for savings, accumulation, and the relationship with the broader capital market. Traditionally, co-operatives have eschewed external finance and have been entirely owned by their workers as a collective group. The labour force or workers, known as members, have an equal share in the profits and equal voting rights. However, since ownership is collective, workers have no individual rights over the assets, not being required to put up a (significant) stake on entry nor being able to withdraw their fraction of accumulated saving on departure. An alternative model, more prevalent in the United States, has co-operatives owned by their members individually, so, while decision-making is on the basis of one member one vote, ownership stakes are transferable via the market-place. The distribution of shares is not necessarily egalitarian in these firms.

Both such forms exist within the United Kingdom. Most contemporary co-operatives follow the Industrial Common Ownership Movement (ICOM) Model Rules, which proscribe individual ownership. Individuals put up, say, £1.00 to join the co-operative, and no additional equity finance is allowed. All remaining assets are held collectively, and the firm must grow by loans from the bank or by plough-back of profit. If the co-operative ceases trading, members have no individual claims on the residual net assets. In contrast, the older British Co-operative Producer Federation (CPF) Model Rules fix no limit on individual shareholding, so that members can have a personal claim on a significant proportion of the firm's net assets. Accumulated capital can, therefore, be either individually or collectively owned, the former via the issue to members of additional shares. If the co-operative ceases trading, net assets are disposed of in proportion to capital holdings.

For both individually and collectively owned co-operatives, the return to capital is assigned to the labour force, and usually paid out in incomes or capital gains. Assuming no external finance, neither pays a scarcity-reflecting price for the capital that it uses in production. As we shall see below, this can lead to distortions in the allocation of resources and in the capital-accumulation process.

An important alternative is for the ownership of capital to be external to the co-operative. Worker-members borrow all their capital, and pay a market price for it. There are no examples of organizations of this sort in Western economies. However, in the Yugoslav system of workers' self-management, ownership is *social*. Workers are granted the right to use the capital, to extend it, and to adapt it. They earn their incomes as the fruits of it. However, they do *not* own it, and are *not* permitted to sell it off or run it down. The capital stock is owned collectively by the society and is merely administered by particular groups of workers. In principle, therefore, labour hires capital under Yugoslav self-management, without the rights and returns of the two factors becoming confused. In practice, however, the Yugoslavs have never charged firms the full scarcity price for capital, so that labour has been able to appropriate some of the fruits of capital.

A second major issue concerns the role of non-members within the co-operative. On the ownership side, the question is whether non-workers can be involved in the decision-making process, as, for instance, would happen if the outside ownership of equity gave non-workers voting shares. Such arrangements may ease financial pressures by widening the resource pool from which the co-operative can draw to finance production and investment, but external control is thought to undermine co-operative principles. Attempts to embed producer co-operatives within a broader planning system might also imply consumer or state representatives on enterprise management boards, and these are likely to raise similar control problems. Henceforth we shall assume that control remains vested solely in the hands of employees. By implication, if the co-operative requires external funding, it must be debt rather than equity finance.

On the labour side, the crucial question concerns non-members: should the co-operative use workers who are not involved in decision-making in the production process? One can see arguments for permitting the use of some hired workers, for example in sectors where the activity is seasonal, so that the number of hands needed at peak times far exceeds the normal establishment. Co-operatives have tended to emerge in such areas—shops, agricultural work, or forestry. The largest US co-operative sector, in the plywood industry, relies heavily on hired labour at peak times. On the other hand, it is argued that the use of hired labour runs counter to co-operative principles, allowing one group of workers to make decisions for another. In the United Kingdom, CPF rules permit the use of non-member workers while ICOM rules do not.

THE PROBLEMS OF WORKERS' CO-OPERATIVES

Despite the numerous attractions of workers' self-management, producer co-operatives have been treated with mistrust or disdain by many on the Left. This is partly because democracy in the workplace is inconsistent with autocratic planning. But there are also real worries about the inherent inefficiencies of such organizations, articulated recently by economists such as

Vanek (1970) and Bonin and Putterman (1987), who point to three broad areas of concern: the responsiveness of co-operatives to market forces, their capacity to invest and grow, and their ability to survive as productive organizations in the long term— the problem of degeneration.

Commencing with the employment and output decisions of producer co-operatives, the key result of economic theory is that such organizations *restrict employment* relative to their capitalist counterparts. There are several dimensions to the underlying logic. If we assume co-operatives to be primarily interested in their members' incomes, profitability implies that pay for members exceeds what would be received in a comparable firm which was capitalist. However, for the firm, wages represent a cost; the higher the wages, the fewer workers the organization will wish to employ. For, given market circumstances, a profitable co-operative will employ fewer workers than its capitalist counterpart, thereby reducing its output below the level that would otherwise pertain. Putting the argument another way, in order to raise incomes, co-operatives will try to raise labour productivity. This will lead them to increase the amount of capital used per employee, implying greater incomes but less employment than would hold in competitive capitalism.

Employment restrictiveness can lead to serious problems in adjustment to economic changes. Capitalist firms are thought to be highly responsive to changes in market conditions, with production increasing to match both increases in demand or declines in input costs. Co-operatives, in contrast, respond far less. For example, an increase in demand always brings forth a smaller response than would be forthcoming in a comparable capitalist organization, because the increased price raises the net revenue of the firm and therefore members' incomes once non-labour costs have been deducted. The co-operative will not take on additional members, because the higher pay would be diluted: 'more ways to cut the cake'. The improved market conditions also offer the opportunity to increase the capital intensity of production and raise labour productivity and thereby incomes. The total response may ultimately be as great as in capitalist firms, once the new capital comes on stream, but the adjustment will be much slower.

More generally, it can be argued that the goals of the co-operative must be first and foremost the collective welfare of the *membership*. The co-operative may have broader social objectives, but these enter because they are desired by particular members rather than because they are inherent goals of the organization. Improved market conditions permit the co-operative to gratify more fully these goals—pay, conditions, hours of work, size of the collective, and so forth—but such objectives are only indirectly satisfied by changes in production. To take an extreme example, consider a co-operative in an unpleasant line of work, where the members seek to reduce their hours of work provided that this does not lead to falls in income. If demand conditions improve, so the co-operative can sell its output at a higher price; the membership may choose to take their higher potential benefits in a shorter worker week, rather than via the increased income that might come from greater output. Thus, to the extent that the gratification of collective preferences conflicts with increased production, the effect on marketable output of an increase in demand will be less than would occur in capitalist firms.

The implication seems to be that co-operatives may not be the best way to organize production in a market economy. Markets are decentralized in order to spread the signals about changing demand, technological, or cost conditions widely amongst a variety of actors. The competition between respondents to market signals is the essence of growth and development. The system is therefore ill served by enterprises which do not react adequately.

But this overstates the problem. The fact that each individual co-operative, mindful of the narrow interests of its members, does not adjust supply sufficiently does not automatically imply inefficiencies in the allocation of resources. The total level of output in the economy depends on the aggregate of production decisions: the choices made by all co-operatives taken together. If existing co-operatives do not react adequately to changes in consumer demand, the resulting misallocations can be tackled by brand new co-operatives. And the system provides economic incentives, in the form of higher incomes, to entrants attracted to meet shortages. Similarly, if input cost reductions or technical advance mean that more of a particular good should be produced, entirely new co-operatives can satisfy the gap in demand left by the inadequate reactions of existing producers. Once again, workers in such co-operatives will earn more than would be available elsewhere in the economy, at least until the shortages are eliminated.

The general point is that, in idealized models of the competitive economy where there are enormous numbers of traders, the restrictiveness of any particular trader does not matter because others will seek to fill any gaps left in the market. Of course, the story becomes more complex when sophisticated production processes are involved, because of the time and expense involved in setting up a business, but the same general principles apply. Provided entry and exit into markets are relatively easy, restrictive behaviour by an individual producer does not really matter. It will be offset by the actions of competitors. In this context, the oft-noted restrictiveness of individual co-operatives fades to secondary importance. The problem instead becomes that they are relatively *hard to form or to close*.

With the creation of co-operatives we immediately run into the 'entrepreneurial problem'. Self-interest entrepreneurs create economic organizations for personal profit. If they spot a viable market niche, they will rarely be happy to share the potential profit around with all their employees via a co-operative. They will prefer the capitalist form, where they can keep the surplus for themselves. Moreover, there are probably additional costs to creating a co-operative, relative to a capitalist firm, if only because the potential collective members have to find each other, rather than just each be hired by the capitalist entrepreneur. These difficulties probably explain the dearth of producer co-operatives in most countries.

Co-operatives also have relatively greater difficulties with respect to bankruptcy and closure. This is because the concept of loss is not well defined in such organizations. Capitalist firms raise revenue by selling their products and incur costs by hiring inputs, including labour, from markets. If the gap between revenue and cost—profit—is persistently negative, the firm will be unable to pay for its inputs and must ultimately close. The resources currently tied up in that line of activity can thereby be reallocated to other, more profitable, uses. In contrast, the surplus of revenue over non-labour costs—net revenue—is the relevant indicator for co-operatives, and is available in its entirety for distribution to the worker-members. If demand conditions deteriorate or non-labour costs size, net revenue is squeezed and labour remuneration must fall. However, when the co-operative is forced to close is up to the members.

Conceptually, the answer is straightforward. When re-muneration falls below what would be earned in the broader labour market, the co-operative is effectively loss-making. If this situation persists, with no reasonable likelihood of its reversal, the co-operative should be wound up and the labour and capital should be shifted to more productive uses. However, the members themselves may be willing to accept very low wages and poor conditions for long periods of time rather than see the organization fail. Then closure will not occur and resources will be frozen in unproductive uses.

From a social point of view, this is not necessarily such a bad thing in a capitalist system, where the alternative to closure may be long-term unemployment. There is a potentially valuable role for co-operatives in a capitalist economy, offering workers the option of continuous employment, perhaps at lower wages, rather than unemployment when conditions are bad. But it must be recognized that social justice for the otherwise unemployed is being bought at the cost of allocative inefficiency. In a market socialist economy, where indicative planning should ensure that new job opportunities emerge in sectors or regions of declining demand (see Chapter 5), the social gain from such efficiency losses would be smaller and probably could not be justified.

In summary, co-operatives have weaker incentives to react to market signals than capitalist firms. The allocative problems that result could in principle be surmounted by entry of new co-operatives in sectors of high demand, and exit of co-operatives in declining industries. But there is little reason to be sanguine about the extent of adjustments from such sources. Co-operative formation is harder than for capitalist firms, because of the suppression of the individual entrepreneur's role, and closure may be prevented because of imprecision about when losses are being made. There will, therefore be serious allocative inefficiencies in a free market system of producer co-operatives.

The second problem area for producer co-operatives is investment. Many argue that, left to themselves, co-operatives tend to invest less than capitalist firms in the same situation.[1] The issue is closely associated with the problem of ownership, and for the discussion which follows we assume that co-operatives use the arrangements currently predominant in the United Kingdom—the assets for the most part being owned collectively by the worker-members. The problem does not arise if the co-operative is owned individually by the members, each retaining a marketable share in the co-operative, as could occur under CPF rules.

To understand the underinvestment problem, first consider a co-operative without recourse to an external capital market, and therefore relying entirely on plough-back for capital accumulation. A capitalist firm in the same situation would undertake any investment project for which the expected rate of return exceeded what the owner could make by leaving the funds in financial assets: the long-term rate of interest. The opportunity cost of funds to each member of the co-operative is of course exactly the same as for the capitalist. However, co-operative members have no individual claims on the collectively owned assets of the organization. Hence the expected return on investment only derives from the expected increase in earnings which will result from the additional capital. This will necessarily

---

[1] The alert reader will have noted that the employment restrictiveness argument, which implies greater capital intensity in co-operatives, contradicts the underinvestment argument, which implies lower capital intensity. However, the former analysis refers to the *demand* for capital on the assumption of perfectly elastic supply and 100 per cent debt financing; the latter to the *supply* of funds on the assumption that investment is self-financed and the enterprise is owned collectively.

be less than the expected rate of return on the capital, because the members lose any rights over the principal invested in the firm. Thus the capitalist receives the return on investment as increased profits, *and* can at any time sell the firm to recoup the sum initially invested. The members receive the return on investment via their incomes, but *cannot* reclaim the sum initially invested because it is owned by the 'collective'. This means that co-operative members will require a greater rate of return from investment projects than capitalist owners in order to recoup the lost principal, which will lead them to invest less than the capitalist firm.

This point can be seen more clearly with the aid of an example. Suppose that technical innovation has led to the development of new production methods in the shoe industry, such that the purchase of new machinery at £1m. would increase profits by £100,000 per year. If the real interest rate is 5 per cent, this would be a profitable way for a capitalist owner to invest £1m. even if he or she were 55 years old and intended to retire in five years. This is because the extra profitability of the company would be reflected in its sale price, so that he or she could earn the £100,000 per year and recoup the £1m. at the point of retirement. Consider the same decision if the shoe company were instead a co-operative, whose 1,000 workers were due to retire in five years. Suppose that the co-operative had made £1m. in distributable profits that year, and the members have to choose between investing in the co-operative or in financial assets. If they do the latter, they each take £1,000 now and earn £50 a year so that they have £1,250 by the end of the period (assuming that they do not invest the interest). If instead they invest in the co-operative, the value of the co-operative's collective assets rises by £1m., which is available for future generations of workers. This generation, however, loses all rights over the investment, and merely earns £100 per annum in increased income for five years, a total of £500. The members would clearly be much better off by taking the profits out of the firm.

As with employment restrictiveness, this characteristic is essentially the consequence of conflict between the individual interests of worker-members and the broader social interest. Individual interests run against saving in the co-operative, even when other sources of finance are limited, because the income forgone over your working life cannot be reclaimed when you leave. The magnitude of the problem, however, is unclear. The wedge that the loss of principal drives between the return on capital and the increase in members' earnings diminishes over time. Hence if co-operative members have a long-term perspective on the firm, this problem will tend to disappear.

But the fact remains that collective ownership gives the co-operative incentives to invest less than their capitalist counterparts. This may be blunted, however, if we allow for the use of external funds to finance investment. In order to invest as much as capitalist firms, co-operatives will have to rely rather more on borrowed finance. But the questions remain whether they will want, and whether they will be able, to borrow enough to offset the shortfall in self-financing. We know from experience that co-operatives are normally worried about permitting 'excessive' external finance because of fears about a resulting loss of control over the future of the firm. If, as a consequence, the collective decides to borrow less rather than more from capital markets, the co-operative will underinvest, despite the availability of external finance. Moreover, banks and other financial institutions typically require a significant degree of self-financing from firms as collateral for the viability of the project. If creditors observe co-operatives unwilling to invest in projects themselves, this may lead them to doubt the wisdom of providing funds externally. As soon as the extent of external financing becomes linked to the degree of plough-back, the tendency for co-operatives to underinvest is reinforced.

More recent economic research has pointed instead to interactions between the co-operative and the wider labour market as the cause of degeneration in successful co-operatives. We have already noted that workers' pay in co-operatives is the income of the firm, net of non-labour costs, per member. Since wages are inherently flexible and loss-making is imprecisely defined, co-operatives are able to maintain employment during recessions by paying their members less than the going rate for the job, with the promise of income recompense during upswings. Organizations which insure employment in this way may be attractive to workers in a capitalist economy when they face the threat of unemployment, in declining sectors, or during downswings in economic activity. Hence we have seen the emergence of co-operative sectors in Western economies during each of the periods of severe depression, such as the 1930s and 1970s.

Yet, for reasons discussed below, few of these firms survive upswings in economic activity. Thus the co-operative sector is relatively smaller during booms. Moreover, almost no co-operatives have grown from being small firms to become large-scale organizations, ensuring the continued existence of a significant and growing co-operative presence of the trade cycle. In practice, co-operatives tend either to remain small and ultimately disappear, or, in growing, to abandon the co-operative structure in favour of the traditional capitalist form (see Estrin and Pérotin, 1987).

The ultimate demise of unsuccessful co-operatives is easy to understand. Co-operatives may be able to survive for some years in unpromising economic environments where capitalist firms would fail, by drawing on the workers' willingness to accept lower wages and perhaps also the dynamism and labour morale unleashed by workers' self-management. However, these motivational effects will not last forever, and, if the co-operative is paying its workers less than the market rate, members will ultimately begin to quit in search of more rewarding employment. Attrition is likely to be concentrated amongst the most marketable, and therefore economically the most important, skill groups, creating a downward spiral of quits and reduced incomes for those who remain. Although the process of closure may be agonizingly long, co-operatives cannot survive forever in loss-making sectors or regions.

Co-operatives may also fail to survive in economic upswings. The argument hinges on the role of hired labour. Consider the case of a co-operative formed in a recession to guarantee employment for its members, at below market rates of pay. As the economy picks up and the incomes of members, which include their share of company profits, begin to rise above those available elsewhere in the economy, we have seen that co-operatives which do not use hired workers will wish to restrict employment below that of capitalist firms. Alternatively, and perhaps more realistically, they will be tempted to recruit hired workers from the general labour market, paying them the going wage, which will be less than their own incomes. There is an inherent incentive for the existing collective to subsitute cheap hired labour for expensive members, in order to raise members' earnings. This is of course a form of labour discrimination, in which non-members are paid less for doing the same work. Hence, as demand increases and the firm grows, it will tend to use hired workers rather than additional members to produce the extra output. Moreover, retiring or departing workers will be replaced by hired workers rather than new members, since this also raises the incomes of the members who remain. As time goes on, the proportion of workers who are members gradually declines until it reaches a level sufficiently low that it is hard to describe the enterprise as a workers' co-operative at all. The discriminatory use of hired labour therefore sows the seeds of internal decay in successful producer co-operatives.

CAN THEY BE MADE TO WORK?

This list of deficiencies would seem to put a nail in the coffin of co-operatives as an organizational form on any significant scale. And the evidence on producer co-operatives in capitalist economies does appear consistent with many of these arguments. For example, in the United Kingdom, as we have seen, co-operatives are typically small, concentrated in skilled labour crafts and trades, under-capitalized, and often experiencing problems of management and control (see Estrin and Pérotin, 1987). There are almost no co-operatives in the heavy industrial sector—steel, chemicals, engineering—anywhere in the world, probably because such activities have large financing requirements. It would seem that co-operatives may be hard put to advance beyond their artisanal enclave, even under market socialism.

But the practical experiences of Mondragon and Yugoslavia warn us against drawing this conclusion too readily. In fact, these problems arise from the particular way that Western co-operatives have been structured, itself largely the consequence of the movement's early history of struggling to survive in a hostile capitalist environment. And as our understanding of co-operatives increases, we are able to devise alternative arrangements which preserve both enterprise-level democracy and economy-wide efficiency.

The kernel of worker's co-operation is democratic control over decision-making. To make this ideal work in a capitalist environment, a number of rules have been developed, in particular with regard to ownership. It is these which restrict the potential for co-operatives as an alternative organizational form upon which to base an economic system. And it is these which must be abandoned for market socialism.

An illustration of the sensitivity of co-operatives' performance to variations in legal form concerns degeneration. The idea that co-operatives will degenerate into capitalist firms hinges on the presumption that worker-members can discriminate against non-members in terms of pay. The obvious legal solution is to make such practices illegal, by proscribing the use of the hired labour. For example, one could follow the approach of Mondragon, making membership a condition of employment in the firms. Similar arrangements apply in Yugoslavia and in the ICOM co-operatives in the United Kingdom.

A more subtle solution has been developed in the large successful co-operative sectors of France and Italy. Both permit the use of hired labour, and indeed up to 50 per cent of workers in French construction co-operatives are non-members. However, the rules require that both members and non-members of the co-operative receive a share of the profits. Hence the extent of pay discrimination is strictly limited. Moreover, in both countries there is the rule that non-member workers must always be admitted to membership status if they so desire. If members attempt to discriminate against non-members, the latter need merely become members themselves to sidestep the problem. Free admission rules of this sort are one way that market socialists could rig the market mechanism to enhance the survival and efficiency of producer co-operatives.

A thornier problem is the collective ownership structure of most contemporary producer co-operatives. One can see the attractiveness of common ownership for the labour movement in a capitalist economy. There is perhaps something socialist about substituting communal for individual ownership arrangements among a group of workers. But when such a structure is replicated throughout the economy, the problems that we have analysed will ultimately emerge. This is because there is nothing to stop each of these groups of workers acting selfishly with respect to the broader society. An economy of this sort is workers' capitalism, not socialism, with capitalists replaced by selfish worker-owners.

Collective ownership is defined *within* the co-operative—an island of socialism in a hostile capitalist environment. But if such arrangements are extended throughout the economy, they undermine the possibility of achieving socialism by maintaining private, though group rather than individual, control over the means of production. The concept of collective ownership must therefore be extended for the socialist environment, to preclude any direct ownership or control by workers of the machines upon which they work. Ownership of co-operatives in a market socialist economy must therefore be *social*, in the sense defined earlier.

It is important to stress that we do not lose the attractive features of co-operatives by such arrangements. Rather we distil their essential characteristic for a different environment. Contemporary co-operatives satisfy socialist aspirations in two ways: on a micro-scale, by the negation of private in favour of collective property, and more generally by embodying democratic decision-making in the workplace. Socialism as a system is in large part about achieving the former aspiration on a macro-scale—by public ownership or by highly egalitarian distribution policies (see Chapters 4 and 6). In a *socialist* economy, the first aspiration is resolved at the level of the economy. It is only the latter which is relevant in organizing the workplace.

The first fundamental principle of self-management in a market socialist economy is that the ownership of financial capital should be separated from the control of production. Enterprises should be run by their labour forces democratically: for example, via general assemblies of the labour force, or via elected employee representatives running workers' councils and management boards. Decision-making will be collective and the labour force has assumed one element of the entrepreneurial function: the right to the residual surpluses (profit from trading after all inputs have been paid for). However, these enterprises should not be owned by their workers. Capital should be treated as an input like any other, borrowed from specialized lending institutions and paid for at market rates. Workers should control the firm democratically but not own it.

Arrangements of this form resolve some, although not all, of the problems raised above. In particular, they eradicate the tendency of producer co-operatives to underinvest. As far as the enterprise is concerned, all finance is external and the capital stock is being hired at a market clearing rate. There is, therefore, no wedge driven between return on capital and the increases in earnings by a loss of principal; workers do not put up any of the funds themselves. The demand for capital is therefore unconstrained by the internal supply of finance, and co-operatives will invest to the point where new equipment raises revenue by as much as the cost of borrowing it. This replicates the conditions for the capitalist firm and ensures an equivalent value of investment.

Social ownership cannot, however, do anything about the problems of resource misallocation. Socially owned self-managed firms suffer from employment restrictiveness just as much as their producer co-operative counterparts. But, as we have seen, problems of misallocation by existing co-operatives are of secondary importance, provided the economy itself is competitive. This brings us to the central issue of enterprise formation and closure. Social ownership allows us to create a new institution, specially devised to undertake this crucial function. The second fundamental principle under self-management is the separation of the entrepreneurial function of receiving the residual surplus from that of sponsoring entry into or exit from productive activities.

Risk-bearing itself should, therefore, be divided into two categories under self-management, to be borne by two different groups of agents. Risks in production should be borne by the existing labour forces of self-managed firms, to whom will accrue the residual surpluses. There is no reason to believe that a democratically organized group of workers will be particularly bad at arranging current production, particularly if they delegate day-to-day control to professional management. Indeed empirical evidence suggests that there may be a productivity boost from such arrangements (see Estrin, 1985). But the entrepreneurial function of spotting new profitable openings, and transferring resources from low to high productivity uses— the formation of profitable new firms and the closure of loss-making ones—should be vested in a separate institution. In this way, the entrepreneurial deficiencies of co-operatives can be filled by alternative market-orientated institutions.

There are several alternatives open to us at this point. The new entrepreneurial institutions could be rather like banks, lending financial capital, monitoring the performance of existing co-operatives, and searching for new outlets for their funds. This would resemble the arrangements in Mondragon, where the central co-operative bank—the Caja Laboral Populaire— plays precisely this role. It also fits the 'European banking model', in which the monitoring and entrepreneurial functions are largely exercised via a centralized banking system rather than through the stock markets of Britain or the United States.

However, market socialists have serious reservations about placing entrepreneurial tasks in centralized, bureaucratic, and public hands (see Chapter 5). One motive for writing this book was to persuade readers of the serious drawbacks to centralized allocation. It would, therefore, be ironic if the only way that democracy in the workplace could be achieved was via central control over key aspects of investment and resource allocation.

Fortunately, an alternative approach is available, which is more in tune with the decentralizing theme of this book. The various entrepreneurial functions discussed above could be vested in a number of competing holding companies, whose primary task would be to manage social capital. Collectively, these holding companies would own all the productive equipment in the economy, and lend it to producer co-operatives at the market rate of interest. We would require many such firms to ensure that the market for social funds was competitive. Strict anti-trust legislation would also be required, therefore, to ensure that individual holding companies never acquired excessive market shares in social capital, either within sectors, within regions, or in the economy as a whole.

I envisage these holding companies as first and foremost profit-maximizing institutions. Their liabilities would be the funds lent to them by their owners and depositors. Their assets would be the social capital that they have lent to productive self-managed firms. Their income would be derived from the interest earned from this capital, and obviously would increase with the size of the asset base. The holding companies would therefore seek continuously to increase the volume of the social capital that they were lending. There are a number of ways in which they could do this. First, they could meet the new investment demands of existing self-managed firms. As we have seen, these are likely to be modest relative to what the market will bear. Moreover, they could attempt to induce firms borrowing from another holding company to switch their debt to them. Perhaps most significantly, the holding companies would be empowered to create entirely new self-managed firms in lines of activity which they considered to be promising. Their role would include research and development, product innovation, market research, and, of course, finance.

The precise relationship between the holding companies and their new client self-managed firms would be highly sensitive. At some point, the task of enterprise formation would have to be defined as complete, so that organizing production could commence. At this moment, the control of the new organization would have to be transferred from the holding company to the production unit's labour force. The holding company would have to establish the product niche, endow the new firm with capital, and hire an initial labour force. However, once production had commenced, the firm would become entirely self-governing, to the extent that, if the labour force so desired, the initial holding company could be paid off and the debt on social capital transferred to a competitor.

The balance of authority between the holding companies and their client enterprises is thrown into sharper relief by the issue of closure. The holding company would have to have the capacity to transfer social capital out of low productivity uses, even against the declared opposition of the workers' council in the firm. Under capitalism, this would be easy; the holding company would close the client when it was unable to pay the rate of return on capital. We have seen that a problem may arise under self-management because the workers could choose to pay the going interest rate on capital, and bear the losses by reducing their pay.

This may make sense in the short run, but preserving jobs at the expense of pay is not a long-run solution. Rather, the holding companies should invest in new activities, creating new employment for the labour force in more profitable lines of production. But for the holding company to intervene in this way it would need to be able to act, not merely on the basis of payment or non-payment of the interest on social capital, but in response to the level of wages in the firm. The guiding principle should be that the holding company could intervene either when the self-managed firm could not pay its social capital debts, or when pay fell below some centrally determined minimum wage. Even then, it would be crucial for competition in the capital market that the self-managed firm in question had the right to transfer its debt to another holding company, who might treat it more leniently. Moreover, at this point of fore-closure the holding company would not have to bankrupt the self-managed firm: it might instead choose to put in new management and capital, but retain broadly the same labour force and productive activity. But, once the self-managed firm was 'loss-making' in the sense defined above, all entrepreneurial functions would return to the holding company until the organization was once again 'in the black'.

Given the crucial role to be played in the self-managed economy by these holding companies, their ownership and control are clearly issues of considerable significance. The issue of ownership in a market social economy is far broader than can be properly covered here (see Chapters 4 and 5), but there are three broad alternatives—public ownership, private equity ownership, and private debenture ownership. The choice between public and private ownership hinges on other distributive arrangements within the economy. It would clearly be inappropriate for a market socialist economy to permit direct private ownership of the means of production on a significant scale. However, the holding companies represent *financial* rather than physical capital, and their power over the production process is strictly limited. Moreover, it is hard to imagine that the competitive entrepreneurial function required of these organizations could be adequately undertaken by bureaucratic agencies of the state. All this points to private ownership of the holding companies, with the strict proviso that fundamental redistributive policies have already been executed.

The choice between equity and debenture ownership brings us to the issue of control. It has been argued that economy-wide self-management requires external rather than internal ownership. If one wanted to follow that route, this suggests that the holding companies should be owned by debenture stock, and controlled by their work-forces. However, these institutions have been devised precisely because of serious reservations about the capacities of self-managed firms in the entrepreneurial field. Conflicts of interest between the worker-member and, for example, profitability might lead the holding companies to be insufficiently attuned to market signals. This is worrying because the efficiency of the economy depends, in large part, on energy, drive, and entrepreneurship in these organizations. At least in the first instance, it might, therefore, seem unwise to extend full self-management to the holding companies. It is instead feasible that self-managed firms themselves might become shareholders in the holding companies, creating a circularity of ownership and control reminiscent of the 'second degree' co-operative of Mondragon. This would be the most attractive solution from my point of view.

Other shareholders could include private individuals, workers, and the government, each of which might be represented on management boards. It is not necessarily the case, under arrangements of this sort, that the *entire* liability base of the holding company be financed by privately owned equity. One might more realistically envisage that the state and private individuals also lend their savings, to be passed on via the holding companies to producers as social capital.

CONCLUSION: ILLUSTRATIONS OF THE TRANSITION

It might be useful to conclude with some brief illustrations of how one might introduce this sort of self-management. I shall consider two examples, the second more decentralized than the first. It should be stressed, however, that the guiding principles presented above are consistent with other combinations of ownership, control, and industrial-structure.

Suppose a market socialist government were elected to office, with an unambiguous mandate to transform relations in the production sphere. All productive enterprises would have to be transformed into self-managed firms and a system of holding companies created to administer the social capital. A relatively centralized approach would be for the state to introduce mandatory workers' control of all productive enterprises above a certain size (to exclude small family businesses and so forth), say fifty employees. Smaller companies could also choose to become self-managed if they so desired, but there would be no compulsion.

Precise arrangements for the democratic control of enterprises by their work-forces would be left to the workers to decide, although the state could provide a series of alternative model rules, and would of course proscribe the use of hired labour. With regard to ownership, the state would transform all publicly and privately held equity into debenture stock, upon which the firms would have to pay the going interest rate. At the same time, the authorities would have to create a number of new holding companies, to each of which would be entrusted certain assets in the national portfolio. Since the state has the task of creating the holding companies, it might choose to retain ownership itself, and would therefore transform individuals' existing equity in various companies into gilts. Under a scheme of this sort, the internal structure of productive enterprises would remain largely unchanged, although of course their system of control would alter. However, an entirely new state-owned capital market would have to be created.

Alternatively,[2] one could build on existing institutions for the holding companies. Thus, if one kept ownership based on equity, current publicly quoted firms could become the holding companies. Their ownership arrangements could remain largely unchanged and the head office would retain its central allocative and monitoring role. However, the various productives and subsidiaries would gain their independence from head office, each being transformed into self-managed enterprises. Each plant or division would therefore owe head office, henceforth their holding company, the value of their productive assets, upon which they would pay the market interest rate. They would then be free to organize production democratically, in any way that they saw fit. For example, consider the case of ICI. Market socialist legislation would transform head office into the holding company, and give decision-making autonomy, on democratic lines, to each plant. Shareholders, not necessarily private individuals however, would continue to own ICI Holdings, but not the productive plants. A variant of this proposal would allow the individual plants jointly to buy the shares of ICI Holdings, presumably in a highly leveraged worker–management buy-out.

---

[2] This proposal was originally suggested by David Winter.

8

Markets, Welfare, and Equality

Julian Le Grand

IN Britain, the welfare state is the largest area of non-market activity outside the family. Some welfare services, such as income support, unemployment insurance, and health care, are provided and financed almost entirely by the state. In others the state operates in conjunction with sizeable private and voluntary sectors: for instance, education, housing, old age pensions, and social care (the care of children, the elderly, and other dependants). In all these areas the state plays a number of roles: as provider, as a source of finance, and as a regulator. It provides services through its own agencies; it subsidizes both its own activities and those in the private and voluntary sector, either directly or through the tax system; and it regulates private and public providers.

For most of the post-war period, although there was much discussion of reform within the system, there was little criticism of the principle of state involvement in welfare. But in recent years even this has been under attack. Philosophers and economists from the New Right have accused the welfare state of inefficiency: of wasting resources on excessive administration, and of unresponsiveness to the real needs and wants of those whose interests it is ostensibly trying to serve. These inefficiencies arise, they argue, because welfare bureaucracies are immune from competition and are therefore run primarily in the self-interest of their employees: the bureaucrats, professionals, and other workers who staff them. Moreover, the welfare state is supposed to create dependency through undermining the incentives to work and to save of its beneficiaries; it also

I am grateful to Nicholas Barr, David Green, and John Hills for helpful discussions of the material in this chapter.

undermines individual conscience through encouraging people to look to the government to take care of the needy.

The New Right thinkers have also seized on the work of more sympathetic critics of the welfare state that shows it not to have achieved full equality, neither within key welfare areas, such as education and health, nor within the wider society. This, argues the New Right, shows that the welfare state has failed even in terms of one of its own priorities: that of promoting greater equality.

Now many of these criticisms are greatly exaggerated. Administrative costs take up a far lower proportion of National Health Service expenditure, for instance, than for comparable private systems; the same is true of state pension schemes compared with private ones (see OECD, 1977; TUC, 1985). There are many dedicated professionals and others working in the social services whose prime concern is with the welfare of their clients. The evidence concerning the welfare state's impact on individual initiative and incentives is spotty, to say the least, and is certainly inadequate to support the more extravagant claims of the critics (Danziger, Haveman, and Plotnick, 1981; Munnell, 1986). And, however inegalitarian they may be in certain areas, the developed welfare states almost certainly have more overall equality and less poverty than societies with more limited welfare policies.

But to say that all is not as bad as has been made out is not to say that all is well. There are undoubted inefficiencies in the operations of the big welfare bureaucracies. Many welfare providers do operate in a way that suggests they are putting their own interests above those of their clients. Although there is little evidence of substantial disincentive effects, there is a widespread perception that such effects exist—a perception that acts as a powerful barrier against increasing the resources going into state welfare even when the latter appears to be significantly underfunded. And there do remain considerable inequalities, both in key areas of welfare provision (Le Grand, 1982; Goodin and Le Grand, 1987) and in the wider society (Rentoul, 1987; Stark, 1988).

Now, as has been argued extensively elsewhere in this book, under certain circumstances competitive markets can be highly efficient. Competitive agencies will economize on resources, including administrative ones. Moreover, by their very nature they are likely to be responsive to their users. Markets reward providers who are sensitive to the wants of their consumers; they penalize those who, at least in the eyes of consumers, provide unsatisfactory service. So one solution to service inefficiency and unresponsiveness might be, as indeed the New Right argues, to introduce market elements into welfare provision—so long as the necessary conditions are met.

But are they met? Even if they are, given the perceived tendencies of markets to exacerbate inequalities, might not market-orientated welfare reform improve welfare provision in one respect—that of efficiency and responsiveness—while simultaneously making things yet worse in another—inequality? Or is it possible to devise market-type welfare systems that create greater efficiency and responsiveness *and* redistribute command over resources from rich to poor? What of the impact of such changes on incentives to work and save? It is to these questions that this chapter is addressed.

Given the breadth and depth of the issues involved, inevitably what follows has had to be selective. The next section discusses the reasons why the simplest market solution of all—full-scale privatization— is not appropriate for most areas of welfare provision. There follows an examination of two proposals for welfare reform that fall short of full privatization but none the less contain substantial market-type elements: vouchers, with specific reference to education, and tax-related charges or user taxes. Finally, I discuss the relationship between redistributive policies in general and overall economic equality, focusing on the specific issue of the use of taxes and transfers to redistribute wealth.

PRIVATIZATION

I begin with the most extreme market alternative to existing welfare arrangements: full-scale privatization. Applied across the main areas of welfare provision, this would involve: private schools, universities, old people's and children's homes, and hospitals; all housing privately owned or rented; doctors and, where appropriate, social workers charging for their services; private insurance companies providing cover for medical care expenses and for loss of income due to unemployment or sickness; and private pension plans meeting the needs of the elderly. Income support for the destitute would be provided through the operations of private charity.

The flaws in such a privatized vision are well known. First, many areas of welfare provision confer benefits to others as well as to the immediate user, benefits that would not be taken into account in a private market. Immunization is an obvious example: if individuals get themselves immunized against a particular infectious disease, this benefits not only them but also, through reducing the spread of infection, everyone with whom they come into contact. Services that confer these 'external' benefits would be underprovided in a completely privatized market. Second, users of welfare services often have insufficient information to make properly informed decisions in the market—a point discussed in more detail below. Third, there are technical problems specific to certain welfare areas, such as moral hazard and adverse selection in insurance for medical care and unemployment, which will prevent private markets from operating efficiently in those areas (Le Grand and Robinson, 1984*a*; Barr, 1987). Fourth, and perhaps most fundamental, the outcome is likely in most cases to be yet more inegalitarian than the existing welfare state, with the distribution of medical care, education, housing, social care, and social insurance being determined primarily by the distribution of market incomes. Private charity is unlikely to remedy this situation, because of its patchiness and sole reliance on the goodwill of the better-off. It may also suffer from the so-called free-rider problem, whereby everyone refrains from making gifts in the hope that others will do so first, thus making further donations unnecessary.

But, whatever else might be said about it, it has to be acknowledged that under this scenario welfare providers would almost certainly be more responsive to their users and clients than they are now. Users of a service who were unhappy with the service offered would not have to suppress their dissatisfaction and put up with the bad service; they could simply go elsewhere. Rude or insensitive providers would lose business to those who were more helpful and considerate. Incompetent bureaucracies would suffer relative to competent ones; if incompetence accompanies size, then, other things being equal, smaller organizations would succeed where bigger ones fail.

Paradoxically, some would claim that this apparent advantage of markets is actually their principal weakness—at least in the welfare field. It presupposes that the consumer always knows best, but in practice, as mentioned above, 'consumers' of welfare services are often ill-informed about those services. Welfare providers generally have access to a range of skills and information way beyond that of any potential client. Under a market system of welfare provision, they have an incentive to exploit the relative ignorance of their clients through providing poor quality service or through providing unnecessary services simply in order to raise the providers' incomes. As a result, so far from promoting efficiency, markets are likely to be inefficient, wasting resources and providing services that do not properly accord with the needs of welfare users.

Further, there is another possibility of exploitation that may arise with privatized welfare: that of families exploiting their own members. In many areas of welfare it is not the clients themselves who make the relevant market decisions; it is someone else within (or occasionally outside) their immediate family. One person, often the husband, commonly decides upon the basic allocation of the family finances. Parents make decisions concerning their children's education. Later in the life cycle, those same children may have to make decisions on behalf of their elderly, confused parents. In such cases there is no guarantee that, in making those decisions, the decision-takers will always operate fully in the interests of the other persons concerned. Husbands may—and often do—put their own financial wants before the needs of the rest of the family. Poor parents take their bright children out of school to put them to work so as to increase the family income. Elderly people are put in a home by their families against their wishes. It is not frivolous to argue that a major role of the welfare state is to protect individuals from their families.

So the full privatization of welfare is not the answer. But some of the efficiency and responsiveness advantages of markets are real. Are there market-type mechanisms short of full-scale privatization that could be introduced into the welfare area that could reap these benefits, without incurring the attendant disadvantages of inequality, of inefficiency arising from ill-informed decisions, and of family exploitation?

VOUCHERS

The most venerable of the market-type ideas for welfare reform is that of education vouchers; according to some accounts, the idea dates back to Tom Paine. It has many variants (see Blaug, 1984), but one version promulgated by reformers from the New Right is as follows.

All state schools are converted into private profit-making institutions. None receives any direct grant from the government. Instead, everyone with a child at school is given a voucher equal in value to the cost of, say, one year's education for that child. This they present to the school of their choice; the school provides the education and presents the voucher to the government, who redeems it for its cash equivalent. If the school wishes, it can charge more for the year's education than the value of the voucher; however, the extra has to be paid by the parents.

Successful schools would attract pupils, vouchers, and therefore funds. Inefficient schools and ones that ignored the wishes of their actual and potential clientele would lose pupils and income; eventually, unless they mended their ways, they would be forced to close down. Efficiency and user responsiveness would thus be ensured. One potential area of family exploitation would be avoided—that of using a state subsidy for purchases other than that intended—since the vouchers could only be used for educational purposes.

Vouchers could be extended up the education ladder. Higher education vouchers could be given to those going on to college and university. Again, these institutions would receive no grant from the government to support their teaching activities; these would be financed from the redemption of vouchers.

The idea can also be applied outside the education area. For example, all the various ways in which housing is subsidized in the United Kingdom, from housing benefit to mortgage-interest tax relief, could be merged into a single housing voucher. Vouchers could also be offered for disability aids, for residential homes, or for other forms of social care. Vouchers could be available to purchase private insurance to cover medical costs, and loss of income due to sickness, old age, and unemployment. Vouchers could even be made available for people to give to the charity of their choice (not as outlandish an idea as it might seem: tax relief on charitable donations—common in many countries—can be viewed as a form of voucher).

Now it is difficult for anyone on the Left to treat the idea of vouchers on its merits, because in recent years the idea has been colonized almost exclusively by the Right. This is a pity, for there seems nothing inherently right-wing or unsocialist in what is perhaps the principal merit of vouchers: that they empower the welfare client. Many of the aspects of vouchers to which the Left would rightly take objection—such as the ability of wealthy parents to top up an education voucher by extra payments—are not essential to the idea. It is perfectly possible to construct voucher schemes that accord in most, if not all, respects with socialist values.

Consider, for instance, the following 'left-wing' education voucher scheme. All schools would be state-owned and operated; perhaps they would be run as teacher co-operatives. Parents would receive vouchers and use them in the way described above. To reduce the possibility of family exploitation, the vouchers might even be given to the children themselves, so long as they were above a certain age (say, over 16). But, to maintain equality, there would be no other way of purchasing education: no private fee-charging schools, and no possibility of topping-up the education voucher by paying more to the school. To allow for the extension of real choice in areas where appropriate schools were not within walking distance, transport vouchers to pay for travel to and from the chosen school would also be available.

Now this would seem to incorporate many of the attractive features of vouchers, while avoiding some of their less desirable aspects. Schools would be forced to take more account of parents' and children's wants than at present. Their responsiveness and efficiency would thus be enhanced. And, through the banning of top-up payments, at least one avenue of inequality would be removed.

There would be problems even with a left-wing voucher scheme of this kind. These should not be used to dismiss the idea out of hand, as many on the Left are inclined to do, but they do need careful attention. The first concerns information. Parents are often poorly informed on education matters; their views tend to be based on their own experience and as a result are commonly out of date and uninformed by any awareness of developments in educational techniques. As was argued earlier, an essential requirement for markets of any kind to work is that users are able properly to assess the merits of the service they are being offered; it could be argued that education is a classic case of this requirement not being met.

This objection undoubtedly has some merit, but none the less has to be treated with care. If parents are ill-informed now, then that could in part be laid at the door of the present system—a system that does not encourage, and indeed on occasion can actively discourage, parents from knowing what actually is going on in the classroom. Parental ignorance can be—and often is—used as a justification for total parental impotence in the face of a corresponding professional omnipotence.

A further problem with the left-wing voucher (and indeed with all vouchers) is that, even if parents were perfectly informed, they still might not make the 'right' decisions about their children's education. This may arise because they put their own interest above those of their children, as we have already discussed. It may also occur because their interpretation of what is in their children's interests does not necessarily coincide with the interests of the wider society. Many people argue that education confers external benefits—benefits to society over and above those accruing to the immediate beneficiaries. Thus it is important to instil a common core of specific values and beliefs in each citizen. There may also be a wider economic interest in the development of certain skills. Complete freedom of choice by parents could result in a divided society and an inefficient economy.

A possible answer to both the problem of parental ignorance and that of external benefits is to have a national curriculum imposed on schools, similar to that currently being implemented in Britain. Professional expertise concerning educational developments would be used in the construction of this curriculum; through the political process, it could also be constructed in such a way as to take account of the wider interests of society.

However, if the national curriculum specified all aspects of educational activity, all schools would be largely identical and the choices that the voucher scheme was supposed to create would be empty. It would be preferable to acknowledge the case for a core of professionally and politically determined uniformity across schools, but to allow a substantial variation by schools, and perhaps local authorities, around the core. More specifically, there should be a relatively small (say, 25 per cent) nationally imposed core curriculum and perhaps an extra local-authority imposed curriculum (up to a further 25 per cent); this would leave at least 50 per cent of the curriculum to be determined by the school and its parents, and therefore allow for the possibility of considerable variation between schools and hence of a real choice between them.

The next problem with vouchers is that they would require careful monitoring to control quality and to ensure that costs do not escalate. Experience with unlimited reimbursement schemes (that is, schemes where the government meets the individual's bill for using a privately provided service) in the United States, such as the Medicare programme for the costs of medical treatment for the elderly, suggests that the costs of such schemes can easily explode. Vouchers would generally not involve unlimited reimbursement; but there would none the less be continuous pressure from providers to raise the value of the basic voucher, pressure that would be reinforced by carefully selected facts and figures demonstrating the inadequacy of whatever was the current value. In the absence of independent assessment, these pressures might be difficult to resist politically.

Another difficulty concerns the fate of institutions that 'fail'. What would happen to the schools that fail to attract vouchers? What would happen to the people who work within them—and to their remaining pupils? Would there be bankruptcy provisions or some other kind of safety net? If so, what form would this take in a system of socially owned institutions? This question is relevant to other areas of market socialism and is discussed elsewhere in this book; suffice it to say here that vouchers raise the issue in an acute form.

But perhaps the major objection to even the left-wing voucher scheme is that, despite the ban on topping-up, the outcome might still be inegalitarian. In particular, it could be argued that vouchers would encourage selectivity—a selectivity that would favour the better off. Successful schools would be swamped by middle-class parents waving their vouchers and demanding admittance. In order to cope with the excess demand, the schools would have to resort to (non-price) selection procedures, such as entrance exams—procedures that in turn would favour the middle class. The remainder of the population would be left with 'sink' schools—schools bereft of bright children (and of pushy middle-class parents), permanently stuck in a mire of low educational standards and uncontrollable classes.

It should be noted that, even if this kind of selection did occur, the outcome in some cases might not be very different from what happens now. Under the present system place of residence rather than examinations is used as the selection process. The middle class move to areas where there are good schools (or lobby vociferously for improvements in the areas in which they already live), thus reinforcing the quality of the services offered and creating a virtuous circle of service improvement, while leaving a vicious circle of decline for poorer areas with ever-poorer facilities. The middle classes have always been adept at manipulating whatever rationing or selection procedures are used to obtain the best service— whether market or non-market.

Moreover, there is a key difference between the present situation and the voucher scheme. The voucher gives actual economic power to all users, *including* those who are poor. Under the scheme proposed here, the purchasing power of the rich and the poor for education would be the same (which could not be said, for instance, about the respective abilities of both rich and poor to move into well-endowed education catchment areas). If one consequence of vouchers is schools that specialize in educating the children of the rich, why should not another consequence be schools that specialize in the challenge of educating children from the poor?

Indeed, this process could be encouraged by modifying the voucher scheme so as to create a discriminatory voucher: one that favoured poor families. They could receive a larger voucher, thus creating a positive incentive for schools to take them on. Schools that contained a reasonable proportion of children from poor families would have more resources per pupil on average than those reserved exclusively for the rich. They would be able to have better premises and equipment and could attract higher quality staff. The outcome would be either selective schools, with those that specialized in the education of the children of the poor being better equipped and staffed than any that specialized in the education of the children of the better off; or, more likely, schools that contained a reasonable mix of children from across the social spectrum. In either case, it would be difficult for a socialist to object.

A difficulty with the discriminatory voucher is that it would be necessary to find some way of identifying poor families; and the conventional way of doing so, means testing, has undesirable features (discussed below). An alternative to a means test would be to use place of residence as the basis for discrimination, with large vouchers being given to families who lived in poorer areas. The wealth of an area could be determined for this purpose by a sample survey of the gross capital value of houses in the area. This would have the advantage of impeding the relatively wealthy from moving into a poor area to benefit from the larger voucher; for, if they did so in any numbers, house prices would rise and the value of the voucher would fall.

A geographically discriminatory education voucher scheme, if introduced in the manner suggested, would not necessarily create greater inequality and would probably create greater responsiveness and efficiency. However, it is equally likely that its introduction would create problems of its own, particularly those of poorly informed parents, of parents making 'mistakes', and of institutions that fail. The likely magnitude of these problems (and whether they would outweigh the benefits) is difficult to gauge in theory. At the time of writing there have been two 'experiments' with vouchers, one a limited trial in Alum Rock, California, and the other a hypothetical exercise undertaken by Kent County Council. Because of their limitations, neither of these is very informative about the kind of issues raised here. What is needed is a serious experiment within an area with a discriminatory voucher scheme; then we could see whether vouchers were dangerous right-wing nonsense or a potentially useful instrument for attaining socialist ends—both in education and elsewhere.

CHARGES AND USER TAXES

The levying of charges for state services currently provided free, or the raising of charges where they already exist, are often suggested as ways of introducing some market considerations into welfare provision. Examples include charging for GP consultations, charging for the 'hotel' component of hospital care, and raising the fees of charged students in higher education to something approaching full cost.

The rationale behind such suggestions is usually to raise further funds for, and to discourage any 'frivolous' use of, the service concerned. But a further justification is to encourage responsiveness: if the charges are in some way related to the income of the providers (for instance, if they are on a fee-for-service basis, rather than simply another source of revenue to go into the general pot), then their introduction may increase the providers' responsiveness to their users' wants.

The problem with the last argument is that, as we have seen earlier, any fee-for-service system operating in an area where clients are poorly informed is likely to lead to wasteful over-use. More generally, a major difficulty with any form of charges is that their introduction at first sight seems unlikely to promote equality; indeed, since a flat-rate charge takes a bigger proportion of income of the poor, the introduction of charges might be thought certain to induce greater inequality.

However, the last argument at least may be too simple. The provision of a service free generally encourages a greater demand for the service than the available supply. This means that other devices have to be used to ration the excess demand— such as waiting or queuing in the case of medical care, or setting tests for entry into higher education. In the case of waiting or queuing, these may act more effectively as a deterrent of use for the rich than for the poor, thus promoting greater equality of use. But other non-price rationing devices, such as performance in examinations or interviews, may favour those from better-off families. In these cases, providing the service free may lower financial barriers to use by the poor but at the same time raise other perhaps even higher non-financial barriers.

Is it possible to devise a system of charging that would discriminate effectively in favour of the poor? The obvious solution is some form of means test, some method of gearing the charge to ability-to-pay. Unfortunately, in Britain at least, means tests have historically been applied in an insensitive, stigmatizing fashion which has often led to low take-up rates for the services concerned as well as frustration and humiliation for those who do apply for the service.

One way round this difficulty is to incorporate the charges into the tax system, via a 'user tax'. This can be illustrated by reference to the idea of a graduate tax (Glennerster, Merrett, and Wilson, 1968; Goodin and Le Grand, 1987, 100–1). This would be a tax set as a proportion of income levied on higher education graduates and collected through the income tax system. The tax rate might vary according to the cost of the education received;

there might also be a limit on the total amount paid. The advantage would be that, unlike a conventional loan, people on low incomes would pay much less than those on high incomes; hence the deterrent effect on taking up low-paid activities would be sharply reduced.

As with vouchers, the idea of user taxes could be applied outside the education area. Indeed in Britain (as in many other countries), social insurance is financed by a kind of user tax: the national insurance contribution (a fraction of which also makes a small contribution to the cost of the National Health Service). This could be adjusted so as better to reflect some of the actual risks involved in different occupations. For instance, those in occupations associated with relatively high mortality, and therefore relatively low claims on state pensions, could pay lower contributions.

Another way of implementing the idea of a user tax is to levy taxes on commodities whose consumption was known to involve the risk of increased medical expenditures, such as cigarettes, alcohol, or motor-cycles. These could have a special tax levied on them, the revenue from which would be used to finance those expenditures—a system that is already being tried in France.

If user taxes were levied directly on incomes, would people be able to opt out of paying the tax through, for instance, contracting into private alternatives? The problem here is what is termed 'adverse selection': it will tend to be the better risks that opt out, thus driving up the costs of the system for those who remain in. To avoid this, the user tax would have to be compulsory for all potential users, regardless of whether or not they engaged in private alternatives.

A final object to the extensive application of user taxes is their impact on a 'marginal tax rate' faced by individuals. They are, as we have seen, a form of means test; and they are subject to the standard problem of means tests that, as people earn more money, they have to pay more for the means-tested service (they face a higher marginal tax rate) and therefore face a reduced incentive to work. However, the force of this objection depends on the context in which the user taxes are introduced. If they are simply added to the present system of income tax, then indeed marginal tax rates will be increased and there may well be undesirable disincentive effects. But the user taxes will generate tax revenue; hence, other things being equal, they will permit a reduction in the rates of the general income tax. There can be no presumption that the introduction of user taxation will raise overall tax rates and hence no presumption that there will necessarily be adverse disincentive effects.

User taxes are essentially income-related prices. As such they are a means of introducing one of the key elements of markets—prices—without creating the standard problems of pricing systems—that they disadvantage the poor. They also have the merit over conventional means tests of being combined with the tax system, thus reducing the stigma and administrative costs normally associated with such tests.

THE REDISTRIBUTION OF RESOURCES

In so far as the discussion of this chapter has been concerned with inequality, it has been with inequalities *within* welfare areas, such as the use of the education and health care systems by rich and poor. What has not been discussed so far is the impact of the welfare state on wider socio-economic inequalities—in particular, on the redistribution of economic resources. This is obviously an area of crucial importance, not only for the arguments of this chapter, but also for those of the book as a whole; for a key part of the latter has been that markets can only achieve socialist ends if there is a greater equality in individual ownership of resources.

The welfare system (taken here to include the tax system as well as welfare expenditures) can affect overall economic inequality in three ways. First, it can reduce inequality in so-called 'human' capital: that is, in individuals' states of health and education. Other things being equal, the healthier or better educated an individual is, the more economic power he or she will command in a market-orientated economy (capitalist or socialist). Second, it can reduce inequality in non-human capital: that is, in the ownership of private wealth, such as property, stocks and shares, etc. Finally, it can intervene in the incomes that people earn from their ownership of both human and non-human capital, through, for instance, income taxation, minimum wage laws, and income support measures.

The ability of the welfare state directly to affect the distribution of human capital and the distribution of income has been the focus of much recent interest and regrettably the space available here does not permit an adequate treatment of all the relevant issues. Instead I shall concentrate upon a topic that has been more neglected in recent years: the redistribution of non-human capital or private wealth through the use of the tax and transfer system.

There are two ways to tax wealth: tax the *holding* of wealth or tax its *transfer*. That is, a tax is levied on an annual basis on the amount of wealth that people hold at the time; or a tax is levied when people transfer their wealth either as a bequest or as a gift.

Several countries, notably Sweden and West Germany, have annual wealth taxes; many others have taxes on particular kinds of wealth, notably property (rates in Britain are—or were—a kind of wealth tax). However, such taxes do have their problems, some of which are weightier than others. First, there is the question of administrative cost. A properly organized wealth tax would require annual valuations of all the assets owned by every household in the land. Some of these would be relatively easy to obtain: stocks and shares, savings accounts, and so on. Others would be much more difficult: unique works of art, country houses. Also, precisely what counted as wealth would have to be decided. Would household furniture and appliances be included? What about pension rights? None of these questions would be easy to answer; and the fact that the assessments have to be carried out at frequent intervals would maximize the opportunity for controversy.

The fact that, as noted above, several countries do operate such taxes must mean that these problems are not insuperable. And indeed, there are ways in which they could be overcome. One, rather appealing method of overcoming the valuation problem, for instance, is for individuals to report their own value for a particular asset and then give the tax authorities the right to buy the asset at that value. Another alternative is to take insurance valuations.

A second problem that might be less weighty than it appears at first sight concerns the impact on wealth accumulation. It might seem obvious that a wealth tax woud reduce the incentive for people to save or to engage in other ways of accumulating wealth. The wealth tax can be escaped, after all, by spending one's savings. But this is too simple an argument. For the tax may encourage some people to save *more* so as to compensate for the effect of the tax on their wealth holdings. So we cannot say *a priori* what the net effect on accumulation will be.

Perhaps the most telling argument against a wealth tax concerns its equity. For it taxes wealth equally, regardless of source. Thus the individual who accumulates wealth through hard work or through a useful invention is taxed on a par with those who inherit a fortune from their family. This could seem unjust: wealth acquired through an individual's own efforts seems to have a rather different status from that acquired through, say, the accident of birth.

This problem, by definition, does not affect the other method of taxing wealth: taxing its transfer. Most countries have some form of taxes on bequests: some extend them to include gifts made in a person's lifetime. These taxes are generally levied at a progressive rate on the amount of the estate or gift that exceeds a (usually very generous) exemption limit.

Transfer taxes have a similar combination of incentive and disincentive effects as wealth taxes. On the one hand, individuals may be discouraged from saving because less of any amount saved can be passed on to their heirs; on the other hand, they may be encouraged to save more so as to compensate for the depredations of the tax. One consequence is unambiguous, however: any reduction in the amounts transferred will encourage the heirs themselves to save.

A disadvantage of transfer taxes based on the size of estates is that they offer no incentive to spread wealth. A kind of transfer tax that does provide this incentive is based not on the overall size of the estate but on the size of the inheritance received by each beneficiary. An ambitious version of this is the *lifetime capital receipts tax*, where the tax paid by an individual on a particular inheritance or gift depends on the total amount of such transfers he or she has received over his or her lifetime (Atkinson, 1972). Those who have received a lot in the past by way of inheritances or gifts pay tax on any new transfer they receive at a higher rate than those who have received relatively little. This has the advantage that it encourages donors to spread their wealth; specifically, it encourages them to give their wealth to those who have benefited little from inheritance in the past. The desire to avoid taxation in any form is a powerful force; the attractive feature of this tax is that this force is harnessed to achieve egalitarian ends.

An administrative problem with the lifetime capital receipts tax is that it requires that records be kept of wealth transfers over all individuals' lifetimes. An alternative that does not have this requirement, yet preserves its egalitarian nature, is to incorporate gifts and legacies into the income tax. The gifts or legacies individuals receive in a year, after all, represent increases in their purchasing power in the same way as their annual wage or salary; why not, therefore, consider them as income and tax them under the present income tax system? Again, the system would incorporate egalitarian incentives; donors would have an incentive to minimize the tax bill on the transfer by spreading it among those with low incomes (rather than, as in the lifetime capital receipts case, to those with previously low inherited wealth).

Whichever method of taxing wealth was chosen, it is likely to be politically unpopular. Political resistance could be reduced, however, by packaging the proposals with some more palatable policy reform. A possibility here that would fit in well with the general aim of wealth redistribution is what might be termed, by analogy with a poll tax, a *poll grant*. The revenue from wealth taxation could be used to finance a universal capital grant that everyone would receive on attaining the age of majority. In that way the accumulated wealth of one generation could be used to fertilize the growth of the next. If the grant was financed by an annual wealth tax, then its amount could vary with the amount of wealth already held by the relevant individuals (or their parents), thus converting the grant into a kind of negative wealth tax. If it came from a lifetime capital receipts tax, it could vary with any wealth already received (the poll grant itself would, of course, be included as capital already received when the tax on any further receipts was assessed).

The combination of a developed system of wealth taxation with a poll grant would go a long way towards the redistribution of resources that is an essential requirement for the kind of egalitarian market socialist economy discussed in this book. However, it is unlikely that it would be sufficient to remove the need for any further redistribution. A world of equal initial resources is not obtainable; but, even if it were, inequalities in income would inevitably arise in the operation of the economic system (of whatever kind), creating poverty, perhaps unacceptable ostentatious consumption, and also, through the influence of income on family circumstances, further inequality in endowments. There would still therefore be a redistributive argument for income taxation and for a social security safety net.

CONCLUDING COMMENTS

Market-orientated reforms of welfare provision, such as voucher and user taxes, particularly if coupled with systems of wealth taxes and poll grants, could make welfare in particular, and the wider society in general, more responsive, more efficient, and more egalitarian. But a word of warning is in order. Social reform is always risky. Proposals for large-scale policy change that sound attractive on paper have an uncanny habit of backfiring in practice. In the case of the ideas discussed here, we have seen that there are many potential problems, some of which, if they prove serious in practice, may outweigh any gains that might otherwise accrue. If any of them are to be tried, then, where practicable, they should be implemented on a small scale and on an experimental basis. Experiments of this kind would have to be carefully monitored; but they are more likely to reveal the true merits and demerits of market-orientated welfare reform than any amount of armchair theorizing.

References

ATKINSON, A. B. (1972), *Unequal Shares*, London: Allen Lane.

BARR, N. (1987), *The Economics of the Welfare State*, London: Weidenfeld and Nicolson.

BLAUG, M. (1984), 'Education Vouchers—it all depends on what you mean', in Le Grand and Robinson (1984b).

BONIN, J. P., and PUTTERMAN, L. (1987), *Economics of Cooperation and the Labour-Managed Economy*, London: Harwood.

BRAVERMAN, H. (1974), *Labour and Monopoly Capital: The Degradation of Work in the Twentieth Century*, London: Monthly Review Press.

BUCHANAN, A. (1985), *Ethics, Efficiency and the Market*, Oxford: Clarendon Press.

CAVE, M. and HARE, P. (1983), *Alternative Approaches to Economic Planning*, London: Macmillan.

COMISSO, E. T. (1979), *Workers Control Under Plan and Market*, New Haven: Yale University Press.

CROSLAND, C. A. R. (1964), *The Future of Socialism*, 2nd edn., London: Jonathan Cape.

DANZINGER, S., HAVEMAN, R., and PLOTNICK, R. (1981), 'How Income Transfers affect Work, Savings and the Income Distribution: A Critical Review', *Journal of Economic Literature*, 19: 975–1028.

ESTRIN, S. (1985), 'The Role of Coops in Employment Creation', *Economic Analysis and Workers' Management*, 19: 345–84.

—— and HOLMES, P. (1983), *French Planning in Theory and Practice*, London: George Allen and Unwin.

—— and PÉROTIN, V. (1987), 'Cooperatives and Participatory Firms in Great Britain', *International Review of Applied Economics*, 1: 152–76.

FORBES, I. (1987) (ed.), *Market Socialism: Whose Choice?*, Fabian Pamphlet No. 516, London: Fabian Society.

GLENNERSTER, H., MERRETT, S., and WILSON, G. (1968), 'A Graduate Tax', *Higher Education Review*, 1: 26–38.

GOODIN, R., and LE GRAND, J. (1987), *Not Only the Poor: The Middle Classes and the Welfare State*, London: George Allen and Unwin.

GOULD, B. (1985), *Socialism and Freedom*, London: Macmillan.

GRAY, J. (1983), 'Positional Goods, Classical Liberalism and the Politicisation of Poverty', in A. Ellis and K. Kumar (eds.), *Dilemmas of Liberal Democracy: Readings in Fred Hirsch's Social Limits to Growth*, London: Tavistock.

214 References

GRAY, J. (1984), *Hayek on Liberty*, Oxford: Blackwell.
HARE, P. (1985), *Planning the British Economy*, London: Macmillan.
HATTERSLEY, R. (1987), *Choose Freedom*, Harmondsworth: Penguin.
HOBSBAWM, E. J. (1968), *Industry and Empire: An Economic History of Britain Since 1750*, London: Weidenfeld, and Nicolson.
HODGSON, G. (1984), *The Democratic Economy: A New Look at Planning, Markets and Power* (Harmondsworth: Penguin).
JACKALL, R., and LEVIN, H. M. (eds.) (1984), *Worker Cooperatives in America*, Berkeley: University of California Press.
LE GRAND, J. (1982), *The Strategy of Equality*, London: George Allen and Unwin.
—— (1984), 'Equity as an Economic Objective', *Journal of Applied Philosophy*, 1, 39–51.
—— and ROBINSON, R. (1984a), *The Economics of Social Problems*, London: Macmillan.
—— and —— (eds.) (1984b), *Privatisation and the Welfare State*, London: George Allen and Unwin.
MACK, J. and LANSLEY, S. (1985), *Poor Britain*, London: George Allen and Unwin.
MARX, K. (1875), *Criticism of the Gotha Programme*.
MEADE, J. (1986), *Alternative Systems of Business Organisation and Workers' Remuneration*, London: George Allen and Unwin.
MILLER, D. (forthcoming), *Market, State, and Community: Theoretical Foundations of Market Socialism*, Oxford: Oxford University Press.
MUNNELL, A. (1986), *Lessons from the Income Maintenance Experiments*, Conference Series No. 30; Federal Reserve Bank of Boston and Brookings Institution.
NOLAN, P., and PAINE, S. (1986), *Rethinking Socialist Economics*, Cambridge: Polity Press.
NOVE, A. (1972), *The Economic History of the Soviet Union*, Harmondsworth: Penguin.
—— (1983), *The Economics of Feasible Socialism*, London: George Allen and Unwin.
NOZICK, R. (1974), *Anarchy, State and Utopia*, Oxford: Blackwell.
NUTI, M. (1986), 'Economic Planning in Market Economies: Scope, Instruments, Institutions', in P. Nolan and S. Paine (eds.), *Rethinking Socialist Economics*, Cambridge: Polity Press.
ORGANIZATION FOR ECONOMIC CO-OPERATION AND DEVELOPMENT (OECD) (1977), *Public Expenditure on Health*, Paris: OECD.
PLANT, R. (1984), *Equality, Markets and the State*, Fabian Pamphlet No. 494, London: Fabian Society.

RAWLS, J. (1972), *A Theory of Justice*, London: Oxford University Press.

RENTOUL, J. (1987), *The Rich get Richer: The Growth of Inequality in Britain in the 1980s*, London: Unwin Hyman.

ROEMER, J. E. (1982), *A General Theory of Exploitation and Class*, Harvard: Harvard University Press.

—— (1986), 'Should Marxists be interested in Exploitation?', in J. E. Roemer (ed.), *Analytical Marxism*, Cambridge, Mass: Cambridge University Press.

RUNCIMAN, W. (1966), *Relative Deprivation and Social Justice*, London: Routledge and Kegan Paul.

RYAN, A. (1984), 'Liberty and Socialism' in B. Pimlott (ed.), *Fabian Essays in Socialist Thought*, London: Heinemann.

SAMUELSON, P. A., and NORDHAUS, W. D. (1980), *Economics*, 12th edn., New York: McGraw-Hill.

SELUCKY, R. (1979), *Marxism, Socialism, Freedom*, London: Macmillan.

STARK, T. (1988), *A New A–Z of Income and Wealth*, London: Fabian Society.

TRADES UNION CONGRESS (TUC) (1977), *Government Review of Social Security: TUC Memorandum of Comments*, London: TUC.

VANEK, J. (1970), *The General Theory of Labor-managed Market Economies*, Ithaca: Cornell University Press.

WALKER, A. and WALKER, C. (1987), *The Growing Divide: A Social Audit 1979–87*, London: Child Poverty Action Group.

WEBB, B. and WEBB, S. (1920), *A Constitution for the Socialist Commonwealth of Great Britain*, London: Longman.

WEITZMAN, M. L. (1984), *The Share Economy*, Harvard: Harvard University Press.

WILES, P. J. D. (1964), *The Political Economy of Communism*, Oxford: Oxford University Press.

WILLIAMSON, O. E. (1986), *Economic Organisation: Firms, Markets and Policy Control*, Brighton: Wheatsheaf.

Index